Modern military matters

Studying and managing the twentieth-century defence heritage in Britain: a discussion document

Modern military matters

Studying and managing the twentieth-century defence heritage in Britain: a discussion document

John Schofield

With contributions by:

Mike Anderton, John Beavis, Jonathan Coad, Wayne Cocroft,
Colin Dobinson, William Foot, Doreen Grove, Vince Holyoak, David Hunt,
Andrew Johnson, Jeremy Lake, Annabel Lawrence, James O'Neill,
Ian Oxley, Bill Reid, Andrew Saunders, and Roger J C Thomas

First published in 2004 by the Council for British Archaeology

British Library Cataloguing in Publication Data
a catalogue record for this book is available from the British Library

The CBA acknowledges with gratitude a grant from English Heritage
towards the publication of this volume.

Cover designed by BP Design, York
Typesetting by Archetype IT Ltd, www.archetype-it.com
Printed by York Publishing Services Ltd, York

ISBN 1-902771-37-0

front cover   *RAF Fylingdales, North Yorkshire. The distinctive 'golf balls' which protected the sensitive tracker equipment of the Ballistic Missile Early Warning System have been demolished and the equipment replaced. (© Crown Copyright NMR, G11722/1)*

back cover   *Pillbox at Kelling, Norfolk. (Photograph: Roger J C Thomas)*

Contents

List of figures and tables

Figures

Tables

List of contributors

Michael J Anderton
(previously) Department for Environment, Food and Rural Affairs

Dr John Beavis
Bournemouth University

Jonathan Coad
English Heritage

Wayne Cocroft
English Heritage

Dr Colin Dobinson
Historical consultant

William Foot
Council for British Archaeology

Doreen Grove
Historic Scotland

Dr Vince Holyoak
English Heritage

Colonel (Ret'd) David Hunt
Consultant

Andrew Johnson
Manx Heritage

Jeremy Lake
English Heritage

Annabel Lawrence
English Heritage

James O'Neill
Environment and Heritage Service, Northern Ireland

Ian Oxley
English Heritage

Bill Reid
Cadw

Andrew Saunders
Fortress Study Group

Dr John Schofield
English Heritage

Roger J C Thomas
English Heritage

Acknowledgements

x

This document was collated by John Schofield, with help from the contributors listed on page ix. Additional comments were provided by Phil Carlisle, Martin Cherry, David Graty, Gillian Sheldrick and Paul Stamper (English Heritage), Antony Firth (Wessex Archaeology) and Nicholas Johnson (Cornwall Archaeology Unit). John Schofield is grateful to all the contributors for their support and cooperation throughout this project. He is especially grateful to Wayne Cocroft, Colin Dobinson and David Hunt for their invaluable contributions in shaping the final script, and resolving the remaining errors, omissions and oversights. Adrian Olivier and Chris Scull of English Heritage are also thanked for their advice on publication and content, as are all members of English Heritage's Military and Naval Strategy Group, and the Defence of Britain Project's Advisory Panel (1994–2002). All comments on this document can be directed to John Schofield at English Heritage [john.schofield@english-heritage.org.uk] who accepts responsibility for any errors it may contain.

Background

The need has been identified for a clear and coherent statement of the state of knowledge and future research priorities relating to the study and management of twentieth-century military remains in Britain. This is a large and diverse subject whose research might variously involve the use of documents, oral history and secondary sources, alongside physical remains in the form of, for example, archaeological and architectural evidence (terrestrial and maritime), wall art and graffiti, and the character or 'personality' of militarised areas. Over the last three decades much valuable work has been undertaken in these related fields by amateur and professional researchers, culminating in national strategic studies such as the Defence of Britain Project and projects commissioned by English Heritage's Monuments Protection Programme (MPP), Historic Scotland and RCAHMS, and other heritage agencies. At the conclusion of these related studies it is timely that we address the state of knowledge, and consider for the first time future research and priorities. Consequently, this report is divided into three sections:

1 resource assessment – reviewing the current state of knowledge
2 research agenda – what gaps exist in our understanding of the subject, and how these research needs might be met
3 priorities for implementing the agenda.

This discussion document and particularly the agenda will be time limited, requiring regular review and modification as the subject matures and develops. Its purpose is to promote dialogue and discussion amongst a wide audience, extend ownership and participation, improve understanding, and to generate ever more refined and intellectually robust research agenda. Yet, this is a document that starts from a position of strength. Much work on modern military archaeology has been completed in the past decade, in the form of coordinated strategic studies, providing a sound basis from which local and more detailed research programmes can proceed. This is not therefore a document that attempts to give focus where focus and coordination were previously lacking. Rather, it attempts to promote targeted academic research, and give a focus to evaluation and archaeological investigations that arise through planning and development control procedures. In other words it is a document that provides context and recommends priorities; affirms the value and cultural benefits of studying modern military sites; and confirms their place alongside other more conventional categories of cultural heritage. As the title states: modern military matters.

Contexte

On a identifié le besoin d'une formulation claire et cohérente concernant l'état des connaissances et les priorités futures des recherches relatives aux vestiges militaires du vingtième siècle en Grande-Bretagne. C'est un sujet à la fois étendu et varié où les recherches pourraient utiliser diverses sources: les documents, l'histoire parlée et les sources secondaires ainsi que les vestiges matériels sous la forme, par exemple, d'indices archéologiques et architecturaux (terrestres et maritimes), les graffitis artistiques militaires et les graffitis, et le caractère ou la 'personnalité' des zones militarisées. Au cours des trois dernières décennies, des chercheurs amateurs et professionnels ont entrepris nombre d'importants travaux dans ces domaines connexes, travaux qui ont mené à des études stratégiques nationales telles que le

« Defence of Britain Project » et à des projets commandités par le « Monuments Protection Programme » [Programme pour la Protection des Monuments] (MMP) de English Heritage, par Historic Scotland et RCAHMS, ainsi que par d'autres agences responsables du patrimoine. A la fin de ces études connexes, il est opportun d'établir l'état des connaissances et, pour la première fois, de prendre en considération les recherches futures et les priorités. Par conséquent, le présent rapport est divisé en trois parties:

1 l'évaluation des ressources – en établissant le bilan de l'état des connaissances à l'heure actuelle
2 un programme de recherches – pour déterminer quelles sont les lacunes dans notre

compréhension du sujet, et comment on pourrait répondre à ces besoins de recherches
3 les priorités pour la mise en œuvre de ce programme

Ce cadre et, en particulier, le programme, seront limités dans le temps, ce qui nécessitera un bilan et une modification à intervalles réguliers dans la mesure où le sujet se développe et prend de l'ampleur. Il s'agit néanmoins d'un cadre de recherches dont la base est solide. De nombreux travaux sur le patrimoine militaire ont été réalisés au cours de la dernière décennie, sous la forme d'études stratégiques coordonnées établissant les ressources, procurant une base solide à partir de laquelle peuvent être poursuivis des programmes de

recherche plus localisés et plus détaillés. Il ne s'agit donc pas d'un cadre visant à apporter la cohérence là où cohérence et coordination font défaut. Il s'agit plutôt d'un cadre s'efforçant de promouvoir des recherches universitaires ciblées et d'apporter la cohérence à l'évaluation et aux enquêtes archéologiques résultant des procédures de contrôle du développement et de l'urbanisme. En d'autres termes, c'est un cadre qui fournit un contexte, exprime des priorités agréées, promeut la valeur et les avantages culturels d'étudier les sites militaires modernes, et confirme leur position aux côtés des catégories plus traditionnelles du patrimoine culturel. Comme le suggère le titre, les sujets militaires modernes sont importants.

Kontext

In den letzten Jahren entstand die Notwendigkeit einen verständlichen und umfassenden Bericht über den aktuellen Wissensstand und zukünftige Forschungsprioritäten im Bezug auf Militärdenkmäler des zwanzigsten Jahrhunderts in Großbritannien anzufertigen. Dies ist ein umfangreiches und diverses Thema, dessen Forschung verschiedenartige Quellen heranzieht, wie zum Beispiel zeitgenössische Zeugnisse, sekundärquellen, sowie auch materielle Überreste, wie zum Beispiel archäologische und architektonische Beweisstücke (an Land und See), Kriegskunst und Graffiti und den Charakter bzw. die „Persönlichkeit" von Militärgebieten. In den letzten drei Jahrzehnten wurde auf diesen Gebieten von Amateuren und Forschern viel nützliche Arbeit geleistet und zu dessen Errungenschaften landesweite strategische Studien gehören, wie zum Beispiel das „Projekt zur Verteidigung Großbritanniens" und Denkmalschutzprogramme die von English Heritage, Historic Scotland und RCAHMS und anderen Denkmalbehörden beauftragt wurden.

Zur Bilanz dieser Studienprogramme ist es daher an der Zeit, daß wir den aktuellen Forschungsstand beurteilen und zum ersten Mal zukünftige Forschungsprogramme und Prioritäten formulieren. Folglich ist dieser Bericht in drei Teile geteilt:

- Quellenbewertung – Bestandsaufnahme des aktuellen Wissensstands
- Forschungsagenda – welche Lücken existieren im Verständnis dieses Fachgebietes und wie können Forschungsbedürfnisse gedeckt werden

- Prioritäten setzen zur Durchsetzung von Forschungsprogrammen

Da es sich um ein neues Fachgebiet handelt, sind die Forschungsprogramme zeitlich begrenzt und müssen regelmäßig auf den neuesten Stand gebracht und modifiziert werden, um somit eine Weiterentwicklung und Reifung des Fachgebiets zu ermöglichen. Dieses Forschungsprogramm geht von einer starken Startposition aus. Viele Forschungsarbeiten, die das moderne militärische Erbe zum Thema haben, wurden im letzten Jahrzehnt abgeschlossen und bestehen aus aufeinander abgestimmten strategischen Studien, die sich auf dem neuesten Quellenstand befinden und somit eine Grundlage für eine Fortsetzung von räumlich begrenzten und detaillierten Forschungsprojekten bilden. Hier soll nicht ein Rahmen geschaffen werden, um Fokus zu erzeugen wo Fokus und Koordination nicht vorhanden waren, sondern es wird versucht einen Rahmen zu schaffen, der zielorientierte akademische Forschung fördert und einen Fokus schafft für Archäologische Gutachten, die aus der Planung und Legislatur der Denkmalpflege hervorgehen. Mit andern Worten, es ist ein Rahmen der Zusammenhänge darstellt und akzeptierte Prioritäten setzt, die Bedeutung und den kulturellen Wert von modernen militärischen Standorten fördert, deren Rang neben herkömmlichen Kulturdenkmälern bekräftigt, und einen Präzedenzfall für archäologische Untersuchungen über andere Aspekte der jüngsten Vergangenheit darstellt. Wie der Titel schon andeutet: Modernes Militär ist wichtig.

Foreword *by George Lambrick*

Coverage of twentieth-century military history on TV and initiatives such as the CBA's Defence of Britain Project reflect the enormous public interest in the history of the rapidly changing world order of the last century.

Modern military matters is a coming of age – a formal recognition that archaeology can contribute much to research into an immensely important part of our recent history. It vividly shows how archaeology is not just concerned with distant epochs and buried remains. Some of what is presented here was both created and became redundant within my lifetime; all of it within the lifetime of many people living in Britain today. It illustrates the process of 'heritage coming over the horizon'. In the 1950s, industrial archaeology (another cause championed by the CBA) began as a crank interest and went on to gain World Heritage status. Twentieth-century military archaeology has won its spurs faster than industrial archaeology – but has been no less reliant on relatively lonely pioneers like Henry Wills, and then the small army of enthusiasts who took up the cause in initiatives like the CBA's Defence of Britain Project. English Heritage and the other state agencies have played a crucial part in recognising the value of the pioneers' work and have initiated and provided support for a very wide range of studies covering different aspects of the subject.

All of this activity is about improving knowledge and understanding as a basis for future management. *Modern military matters* provides a timely and much needed overview of the huge range and breadth of initiatives that have brought this burgeoning field of research to maturity though a well-developed partnership of state, professional and voluntary bodies and individuals. Above all, it points the way forward in a highly flourishing – but still under-appreciated – field of archaeological endeavour. Those of us who are not experts in the field can be grateful to all the contributors for presenting such a wide-ranging subject in such a lucid way.

In many ways, the considered, research-oriented approach of *Modern military matters* contrasts with the more controversial debate over listing of modern buildings. Little of what is dealt with here touches on the issue of architectural design; but there is no question of its great historical, technological and social interest. And as many of the illustrations show, these tangible testimonies to past conflict do have gaunt aesthetic qualities that have come to be appreciated because they are recognised as history, not eyesores.

George Lambrick
Director, CBA

Introduction

Modern military matters covers a wide range of military activity undertaken in Britain during the course of the twentieth century. During the period 1914–89, everyone was in some way touched by the world wars and the subsequent Cold War. A vast area was used for military activity (in February 1944 some 11.5 million acres were under military control for military purposes, 20% of the land area of Britain), and there were unprecedented scales of construction, for example during the RAF Expansion Period in the 1930s, and when the threat of invasion reached its height in summer 1940. Furthermore, the areas around militarised zones were influenced by the presence of military personnel, creating zones whose character or 'personality' has changed irrevocably as a result, even where the presence of the armed forces has been significantly reduced to leave only perhaps some reserve units or cadets. The scope therefore is broadly defined, embracing a wide range of military activity and its influence on the landscape: from coastal fortifications, airfields and radar stations, to army camps, hospitals, training areas, armament factories, bombing decoys, experimental and research sites, aircraft crash sites (on land and at sea), ship and submarine wrecks, and civil defence; and thematically in, for example, the military role of science; and chronologically from the years before the Great War to the fall of the Berlin Wall and beyond.

Our subject, then, is twentieth-century war: a phenomenon distinct from what came before and emerging, already, as different from the wars of today. The dominant technical novelties of World War I were the submarine, air power, chemical warfare, and the tank. The inter-war period saw aircraft technology advance by leaps and bounds, but despite gloomy predictions that the bomber would 'always get through', it also saw the origin of radar, heir to the acoustic early-warning systems pioneered in 1917–18. World War II brought electronic warfare fully into the battle, along with the communications systems which became 'automated data processing' (ADP), eventually supplemented by Information Technology (IT). That war also saw an invasion threat, built around airborne troops and armour, but with the Battle of Britain won – by a system fielding guns, searchlights, decoys, barrage balloons – Britain's war at home became an attritional affair. By then the emerging technologies were the jet engine, the guided missile and the atomic bomb. So they remained, through the Cold War; and with biological and chemical threats bubbling in the background, nuclear weapons completed the sinister triad of 'NBC'. 'Total War' – *the* strategic theme of the twentieth century – had potentially never reached such heights, or depths. Logistic activity has varied to suit this pattern of events, while intelligence effort has multiplied and alliances have come and gone. Above all, perhaps, we are struck by the sheer pace of change.

All of these developments have generated a distinctive military infrastructure, presenting some equally distinctive conservation challenges and stimulating new perceptions and approaches. *Modern military matters* explores some of these, and in particular advocates fresh themes for research, inviting discussion of methods suitable to build upon the solid achievements of the last fifteen years. Our instincts are inclusive and eclectic. It is clearer now than a decade ago that our study demands a multi-disciplinary approach, drawing upon documents, building analysis, archaeology, oral history and other sources to draw a rounded picture of our military past, and the fabric it has left behind. And context is all: historical geographers, social and cultural historians, and students of psychology and literature all have their parts to play.

As has often been said, war (and more specifically

Figure 1 RAF Neatishead, Norfolk. Early 1960s Type 84 radar – this is the last surviving large fixed Cold War radar in England (© English Heritage, AA98/05747)

Total War embracing everything and everywhere, involving entire populations and not just the warriors amongst them) has been a defining characteristic of the twentieth century. There can have been very few people not directly or indirectly affected or influenced by war or conflict somewhere in the world and at some point during the twentieth century. War has shaped the modern world, from the physical scars of – and the preparations for – battles, to states of mind, political and ideological positions, and to the language we use. It has also brought significant scientific developments, which benefit the world in peacetime. The immediacy and the relevance of twentieth-century war has meant its prominence in educational curricula, and its growing influence on cultural tourism. There is a growing demand for informative, factual television documentaries. People want to know more about this period of history and the impact of warfare and militarisation on society, but not just from television, books and the Internet. People want to visit the remaining structures; they have an interest and that interest is burgeoning. The surviving sites are therefore important to satisfy that growing interest and demand, and for reasons of memory, commemoration and sense of place, reflecting key points in the *Power of Place*, (English Heritage 2000), *Force for our Future* (DCMS 2002) and social inclusion agenda. Archaeology and historical research as subjects, whether fieldwork and survey or the analysis of documentary or oral-historical sources, can contribute much to understanding – and thus to commemorating and remembering – twentieth-century warfare, and hopefully learning lessons from it.

The recognition of value, for all of these reasons, has a long history. The significance of artefacts from World War I was recognised even before the end of hostilities, when depots were established in France to collect material for the future Imperial War Museum. Also from soon after the Armistice the first tourists arrived to see the battlefields, some of which were already monumentalised. These early visits were more concerned with commemoration and remembrance than with historical study and

Figure 2 Atomic Weapons Research Establishment, Orford Ness, Suffolk. This building, known as a 'Pagoda', was built during the 1950s for testing the non-nuclear components of Britain's atomic bombs. The National Trust now owns the site (© English Heritage, AA021752)

research, however, reflecting changing perceptions brought by time to modern military remains.

The significance of twentieth-century military monuments was also realised immediately before World War II when the statutory protection of concrete emplacements (pillboxes) from World War I was discussed. Although not protected at that time, their significance as 'ancient monuments' was at least recognised. Research into these surviving structures began in the 1960s–70s, when leisure time was increasingly available, and at a time when much archaeological work was being undertaken by amateur groups and individuals. At this time the archaeology of World War II became the subject of increasingly intensive and sustained effort amongst these amateur archaeologists. Henry Wills' book *Pillboxes*, published in 1985, and his BBC Chronicle Award in 1979, exemplified this period of study.

The late 1980s and in particular the 1990s saw a burgeoning interest in this subject amongst professional archaeologists concomitant with the growth of archaeological resource management: the need to value, prioritise and manage our cultural heritage in an appropriate and sustainable way. This greater awareness was also promoted by the few professional archaeologists involved directly in the subject: Andrew Saunders, in his *Fortress Britain* (1989), for example, referred to twentieth-century remains alongside fortifications of earlier periods.

Figure 3 Anti-invasion defences at Wilsthorpe Cliff, East Yorkshire (Photograph: Roger J C Thomas)

In the 1990s several research projects began documenting twentieth-century military remains, including notably work by the three Royal Commissions in England, Wales and Scotland (eg RCAHMS 1999a and b). In addition, Historic Scotland commissioned regional assessments of the survival of twentieth-century defences, while an audit of surviving remains was also completed for Pembrokeshire (Thomas 1994). One of the first surveys of a modern military site (as opposed to a site picked up in the survey of earlier remains) was at Bowaters Farm, Essex (RCHME 1994a); it was also one of the first World War II sites to be scheduled under the English Heritage Monuments Protection Programme. In 1993, the National Trust took on the management of Orford Ness, Suffolk, an experimental site spanning the Great War to the Cold War, while the national heritage agencies began responding to demand by laying greater emphasis on interpreting recent history at their properties, notably at Dover Castle and Fort George. The Environment and Heritage Service, Northern Ireland are now routinely recording monuments representative of the 'Troubles', prior to their removal as part of the ongoing peace process.

The starting point for studying the military heritage is of course the higher-level strategic policies of successive governments, and these remind us of the importance of alliances, and the resultant diversity of influences upon building and site design. Britain's abiding strategic alliance in the twentieth-century – as now – was transatlantic; American troops were stationed here in the Great War and the Second World War, and then, throughout the Cold War and beyond, as part of the US contribution to NATO. In the two world wars US and other nationals were generally accommodated in fabric of British build and design (bomber airfields are a notable example), but the infrastructure resulting from Britain's NATO membership was shaped by the operational requirements of the Alliance, and built essentially to NATO rather than British standards.

In 1994–95 two national initiatives were launched representing the culmination of effort over a longer period: the Defence of Britain Project and a series of related projects commissioned by English Heritage (*Twentieth-century Fortifications in England*), later extended to cover documented sites in Wales, Scotland and Northern Ireland. Together these studies have begun to document both the scale and extent of militarisation during the twentieth century, as well as recording the survival and condition of World War II sites sixty years on. Work on Cold War monuments has also now been completed. These surveys will be described in more detail in the Assessment that follows, as will some of the locally based surveys for which these national programmes have provided context (eg Nash 2002; Smith 2001). Finally, alongside this greater awareness of twentieth-century military remains, and of their significance and role in contemporary

Figure 4   A Royal Observer Corps post at Watton, Norfolk. To the rear is the early 1950s raised Orlit post for the visual plotting of aircraft movements. In the foreground is the entrance to the late 1950s underground monitoring post, which was designed to detect and monitor the results of any nuclear attack on the country (© Crown copyright. NMR BB98/30036)

society, the National Monument Records (NMR) and locally held Sites and Monuments Records now typically incorporate these sites, embedding them further as part of Britain's cultural heritage.

This document comes at a time when the significance of archaeological remains is better understood and better represented in heritage legislation and non-legislative government advice than ever before. Also the threats to the archaeological resource are better documented (for example through studies such as the Monuments at Risk Survey, and work by the Monuments Protection Programme), but at the same time those threats are increasing. Some military sites in the east are suffering badly from coast erosion, notably coast artillery batteries, radar stations and anti-invasion works from 1940–41. Brownfield sites, including some on the military estate, are under threat given government priorities for future housing requirements and the changing needs of military training. Sites may also be under threat from the Aggregates Levy (from 1 April 2002) which could make many abandoned reinforced concrete military sites potentially attractive sources of aggregate. A significant class of archaeological site – military aircraft crash sites (both on land and underwater) – is under threat from uncontrolled

excavation, with the information retrieved often feeding only into private collections as opposed to public archaeological records (but cf. English Heritage 2002). In general there remains a perception that military sites only constitute ugly, unprepossessing and dangerous structures. Some are all of these, though these considerations are of course irrelevant in assessing significance.

Finally, this framework also comes at a time when war and terrorism once again threaten to disturb the fragile peace that some of us enjoy. It is at times like this that the need to recognise this recent military heritage comes sharply into focus. No part of this heritage should be forgotten or ignored. It has the potency to evoke periods of national unity and achievement, at times – notably in 1940 – against seemingly impossible odds. It also has the capacity to open wounds, such as perhaps does the physical legacy of the Troubles in Northern Ireland or, to look to the heart of Europe, the fall of the Berlin Wall. We can learn from monuments of both kinds and ensure that, however painful their impact may be on some, they or some tangible record of them remain to stimulate reflection and debate, both for ourselves and future generations (for the Berlin context see Dolff-Bonekaemper nd).

Part 1: Assessment: the known resource

Twentieth-century defence studies is a multi-disciplinary field. Significant contributions to understanding this subject have stemmed from: archives and documentary sources; oral history; official histories and other academic writings; buildings recording; historical geography; archaeological survey; and – to a small extent – excavation. The following examples summarise, and are indicative of, some of the projects undertaken in Britain, and reflect the diversity of approaches within this field of study. This framework is a record of work undertaken to date; while several notable projects are described in outline at least, the survey does not claim to be definitive.

Theme 1: The militarised landscape

This theme represents the broadest context for military activity, and is the geographical framework within which all of the other themes are situated. It includes every aspect of militarisation and military influence, from actual sites to the impact of those sites on the natural environment and agricultural systems, to the influence of military personnel on the personality or character of places and landscape. It is all-encompassing in other words and should be regarded as the theme that underlies all others.

The militarised landscape

World War I witnessed a profound change in how the military used the landscape to prepare for and wage war. This change in use created demands for land for camps, training areas (both land and sea), dockyards, new armaments factories, and airfields. This was in addition to the strain being placed on the land to feed

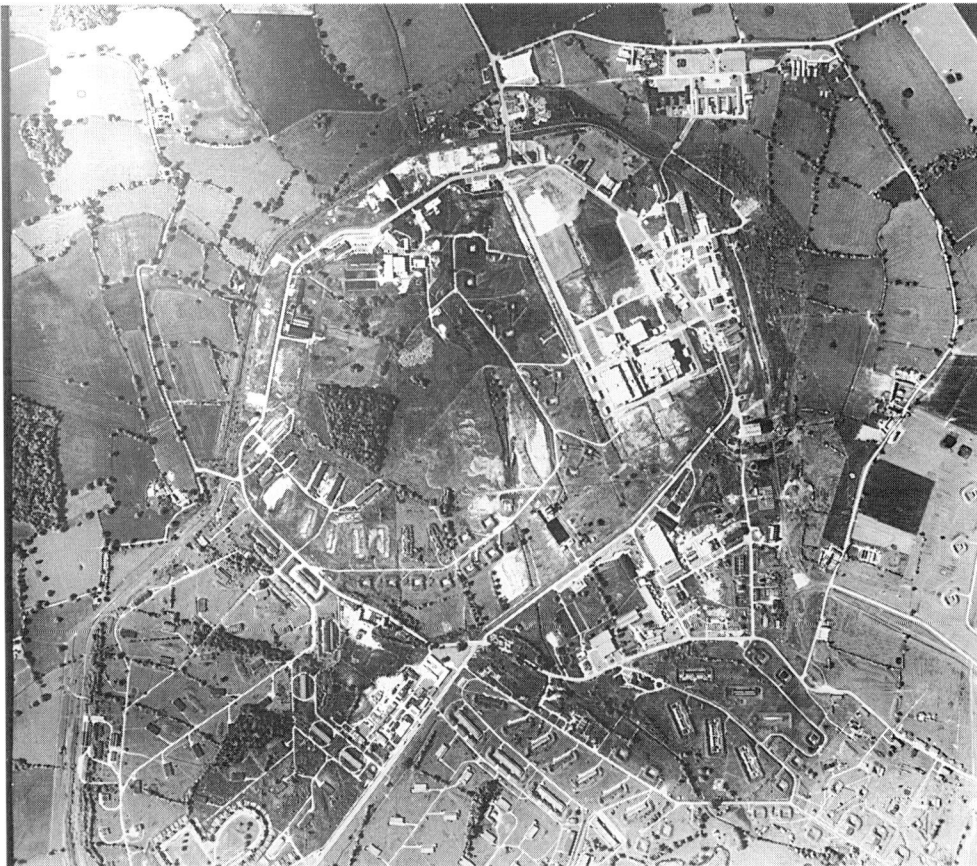

Figure 5 Royal Ordnance Factory Wrexham, Clwyd. This view of a World War II cordite factory illustrates the huge areas of agricultural land appropriated by some military installations. Post-war, many of these factories formed the basis for large trading estates (© MoD Crown Copyright, 58/5171 Frame 0015, F22, 6 June 1962)

Figure 6   A Tobruk shelter at Linney Burrows, Pembrokeshire. This mock-German machine gun post was used to train the 79th Armoured Division in advance of D-Day (Photograph: Roger J C Thomas)

Figure 7   Modern anti-tank missile firing points at the Okehampton military training area, Dartmoor (Photograph: John Schofield)

a country previously heavily reliant on imported foodstuffs.

With World War II, and particularly in the build up to D-Day, much of the landscape was used for military and related purposes. By June 1944, 3.5 million military personnel were sharing the land with the civilian population, and 11.5 million acres were directly or indirectly under military control (20% of the total British land surface). The militarised landscape was acquired and defined through the Defence Regulations, which came into effect under the Emergency Powers (Defence) Act 1939. This included the power to requisition land and evict the civilian inhabitants. The Act was used freely throughout the war. Some one million acres were requisitioned in this way.

Given this intensity of military occupation, it is not surprising that such enormous evidence of wartime construction, land use, and purpose still survives. The physical evidence is all around us, in whatever part of Britain we live, though much also now remains out of sight, buried over the years or submerged on the seabed. Moreover, certain major cities were bombed heavily during World War II, leading to redevelopment programmes in the post-war period (eg in Plymouth and Coventry), contributing another significant dimension to the archaeology of Total War. Every village and town retains memories of its wartime years, whether of the bombing, the presence of troops or the works they left behind. Much of that memory resides with the civilian and military survivors, a fragile resource which will inevitably decline in coming years.

[OBJECTIVES IN PART 2: A, B (ESPECIALLY B14), C3–4, D4, E, F]

Military training

In September 1939, the area of land occupied by the army for all purposes, including training, was 235,000 acres. By February 1944, this had risen to 9.8 million training acres alone. Land was required for infantry and tank training, for beach assault, and for weapon practice. Coastal sites involved not only use of the land but also the seabed designated by navigational exclusion zones. Where the land was selected, often requiring the expulsion of its civilian population, camps for the incoming troops had then to be built. The villages and farms occupied by the civilian population prior to their removal are now part of the military landscape. It is worth noting that the prolonged military occupation of some land has allowed rare flora and fauna to flourish and has protected ancient monuments and landscape from destruction through development and modern ploughing.

A review of the principal army training areas in World War II (embedded within a more general survey of twentieth-century military training establishments) has been produced (Dobinson 2000a). The main ones include:

Redesdale; Otterburn; Fylingdales; Yorkshire Wolds; Stanford/Thetford; Southwold; Dunwich; Orford; Wye; South Downs; Salisbury Plain; areas of Gloucestershire, Oxfordshire, Berkshire and Wiltshire; Slapton Sands; Studland; Lulworth; Dartmoor; Woolacombe; Sennybridge; Gower Peninsula; Castlemartin; Warcop; Ysbyty Ystwyth; Fforest Fawr; Inverary; Tarbat Peninsula; Burghead Bay; and Culbin Sands.

The spatial significance of these training areas has also been assessed, and their iconography considered (Tivers 1999).

In addition, there were hundreds of both permanent and improvised small arms-firing ranges spread widely over the landscape, for Royal Navy, Army, Home Guard, Auxiliary Unit, Royal Air Force, United States Forces, and other unit use. Other training ranges, often on the coast, were those required for bombing and firing practice by the Royal Air Force, the United States Army Air Force (USAAF, but USAF after 1947), and the Fleet Air Arm. In addition to the surrounding camps, the training areas required firing points, stop butts and observation and range control bunkers, examples of which survive. Finally, much wartime training was done under defence regulation 52, not in set areas but over open countryside.

The above listing admittedly represents a diverse set of sites and structures, whose drawing together is arguably justified by their demanding large areas of land – individually in most cases, but collectively without doubt. This criterion alone, perhaps, places them in a different category with respect to the 'militarised landscape' compared to many sites of smaller physical extent. So too does the tendency of many – though not of course all – to be established for long periods of time, and thus to consolidate and entrench their social and landscape effect – to influence whole areas, communities and districts. Beyond these there were of course many other types of sites used for training, urban and rural, not best considered here; the Dobinson (2000a) study lists many and training in general emerges later in the document as a notable area for future research.

[OBJECTIVES: A1–2, A13, B1–2, B7, B9, B12, E4–5, F]

Theme 2: Research and Development and manufacturing

Technology, experimentation and research

In the modern world, war, and the preparations for war, have been amongst the most important stimuli for industrial, technological and scientific advancement. This topic has been widely studied by historians of technology, yet little archaeological research and recording work has been carried out (but cf. Cocroft 2000). In the nineteenth century

Figure 8   Control building at the Missile Test Area, RAF Spadeadam, Cumbria (Photograph: Roger J C Thomas)

most Research and Development work took place within manufacturing establishments, pre-eminently at the Royal Arsenal, Woolwich. Through the twentieth century the State became progressively more involved in scientific research, much of it directed towards military ends, and this was reflected for example in the establishment of the National Physical Laboratory at Teddington in 1902. During the world wars the success of the armed forces became ever more dependent on scientific research (such as the rapid development of the effectiveness of submarine warfare), so much so that World War I, with its demands for explosives, and later poison gas, is sometimes known as the 'Chemists' War'. In World War II, new scientific advances such as radar, the jet engine, rockets, computers and the atomic bomb were applied to warfare leading to this being dubbed the 'Scientists' War'. To meet growing demands, dedicated research establishments, often with highly specialised test equipment became a feature of the defence estate. The importance of these Research and Development establishments grew throughout the Cold War and many were set up to investigate specific problems associated with new technologies, including nuclear weapons and jet propulsion. Orford Ness is one of the better known experimental sites, now owned by the National Trust (Wainwright 1996).

A handlist of experimental sites in England was

drawn up from documents in the National Archives (previously the Public Record Office or PRO) (Dobinson 2000a). In virtually all areas detailed information on the research programmes carried out at particular sites and the structures erected to support those activities is currently lacking, although the records do generally survive. One exception is sites associated with rocket propulsion that were investigated as part of the RCHME explosives industry project (Cocroft 2000). Discounting the post-war structures recorded at the former Royal Gunpowder Factory, Waltham Abbey, Essex (RCHME 1994b), no full archaeological investigation and recording has taken place on a Cold War period research site in the UK, although a project in England involving the rocket testing facility at Spadeadam in Cumbria started in 2003. In addition to the sites listed by Dobinson (op cit), research activities were also undertaken on firing ranges and other defence establishments as well as the premises of defence manufacturers and some academic institutions.

The use of the Colossus computer in code breaking at Bletchley Park in World War II is well known, and in the post-war years the armed forces were early users of Automated Data Processing (ADP), replacing their annual pay roll and inventory control functions with computers. Computers were used also in the Telegraph Automatic Relay Equipment (TARE) to replace the manual tape relay stations in the single service networks. Many buildings associa-

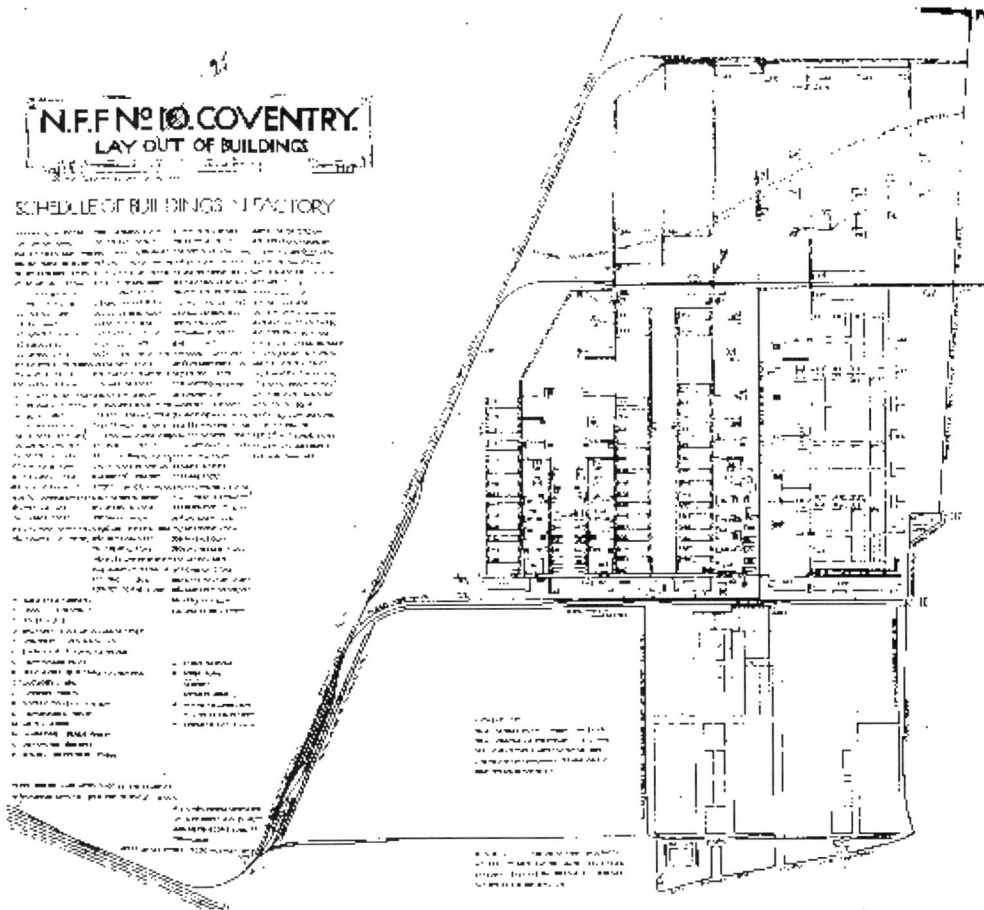

Figure 9   National Filling Factory No 21, Coventry. World War I factories brought to the fore new concepts in factory design, including standardised manufacturing units with production broken down into a series of relatively unskilled tasks. They also brought new standards in welfare, including changing rooms, canteens and medical centres (© The National Archives (previously PRO), MUN4/11220)

ted with these pioneer systems survive. Meanwhile, the armed forces' growing dependency on Information Technology has parallelled the similar trend in wider society; indeed military operations based upon computer networks and their protection have become specialist fields in themselves. Much information on these systems is available in the contemporary defence press.

[OBJECTIVES: A3, A7–9, B1–2, B12, D3–4, E1–2, E4–5, F]

Munitions production

Just as the needs of the armed services have driven scientific research, they have often also been at the forefront of industrial manufacturing techniques, their products often being the most complex of their age. In the nineteenth century exemplary developments include Brunel's innovative block cutting shop at Portsmouth and the adoption of the concepts of mass production and interchangeability by small arms manufacturers. By the end of the century the relationship between industrialists and the military establishment had become so close that the contentious concept of a 'military-industrial complex' had started to emerge. In the twentieth century the scale of required production was so great that the State either closely managed, or assumed direct control of the manufacture of war matériel. Many of the purpose-built factories constructed to meet these needs introduced novel ideas about production methods as well as bringing new standards of welfare facilities for their workforces.

The study of munitions production sites has traditionally been regarded as industrial archaeology. Few national surveys are available detailing individual industries or twentieth-century factory design, an exception being the discussion of explosives manufacture during the world wars in *Dangerous Energy* (Cocroft 2000). Other aspects of the munitions industry in England, such as engineering and chemical and biological related munitions production will be covered in assessments of these industries undertaken by English Heritage's Designation Team. Gretna, in Dumfries and Galloway, is a town built on this industry, and like many others, any study of it should consider the industry's social impact alongside any engineering achievements.

In the early twentieth century most munitions industries were located in the traditional metal working and shipbuilding areas of the country – the Midlands, the north of England and Scotland – as well as the naval dockyards in the south of England. During the 1930s rearmament period, the threat of air attack from the continent led to many new munitions factories being located in the west of England and Wales. The threat also resulted in major efforts to construct underground munitions stores in Britain. In the post-war period there was another marked shift in the geography of the industry as the Cold War encouraged the growth of aerospace, communications and electronics industries. These tended to cluster in the south-east of England, frequently on the fringes of the New Towns within reach of London, and many were clothed in the new factory architecture of the 1950s.

In nearly all areas of munitions manufacture national syntheses and detailed site studies are lacking, although the latter are becoming more common, often in response to redevelopment proposals. Local or regional rapid assessment studies, such as the *Buildings of the Radio Electronics Industry in Essex* (English Heritage 2001), are valuable in highlighting sites worthy of further recording work or conservation.

[OBJECTIVES: A3, B1–2, B12, D3, E2, E4–5, F]

Shipbuilding

British shipbuilders and repair yards played key roles in building and maintaining naval and merchant vessels. In both world wars an average of half a million tons of shipping was being repaired at any one time as the result of enemy action. Although small shipyards, such as those at Cowes and Falmouth, made a significant contribution, the bulk of construction and repair work was done at the main shipbuilding centres – the Mersey, Barrow-in-Furness, the Clyde, Belfast, and the North East. As a general rule, commercial shipyards were unable to afford the substantial buildings favoured by the government dockyards, and with the collapse of British shipbuilding in the 1970s and 1980s, the more ephemeral nature of commercial yards led to rapid site-clearance. One notable exception is the vast submarine-building yard at Barrow-in-Furness. The survival of shipyard archives has also been patchy.

[OBJECTIVES: A3, A10–11, A13, B1–3, B12, D3, E2, E4–5, F]

Design, development and manufacture of armoured fighting vehicles

The tank was a British invention, and Britain led the world in tank design and the exploitation of armour during World War I. After World War II the need for battle tanks arose primarily from the threat posed by the Soviet and East German armies in the NATO Central Region. The Conqueror, Centurion and Chieftain main battle tanks were specifically designed for British forces assigned to NATO in Germany, and were built either at the Royal Ordnance Factory at Leeds or at Vickers. In addition, Britain designed tank guns, engineer vehicles and bridging. There is considerable infrastructure relating to the development, production and testing of armoured vehicles in the UK.

[OBJECTIVES: A3, B1–2, B12, D3, E2, E4–5, F]

Aircraft manufacture

Aircraft manufacture has been an active field of study in recent years, especially among students of aviation and business history, though the full range of sites and structures associated with the industry – let alone their survival – has yet to be assessed (though cf Gillett 1999). Notable examples already identified, however, include the two airship construction sheds at Cardington (one transplanted from Pulham in the 1920s) and listed Grade II*, the Sopwith offices and Hawker workshops at Kingston (Surrey), the Bristol works at Filton – one of the earliest in Britain and distinguished today by the 1936 company offices – and wartime structures at Brooklands.

[OBJECTIVES: A3, B1, B12, D3, E2, E4–5, F]

Theme 3: Infrastructure and support

Naval bases

Great Britain entered the twentieth century with the world's most powerful navy. At that time, some 2.3 % of the male working population was engaged in naval orders, in the royal dockyards, in commercial shipyards and in allied industries. The British merchant fleet similarly dominated imperial and world trade routes; like the Royal Navy, it depended on UK shipyards for the overwhelming majority of its ships. With no serious threat to its supremacy since the end of the Napoleonic Wars in 1815, the Royal Navy's chief role in the nineteenth century had been the protection of trade routes, the policing of Empire and the suppression of the slave trade. It accomplished these tasks using a series of bases, later joined by coaling stations, located throughout the Empire. These overseas bases never rivalled the home dockyards in either scale or complexity; only the latter built the fleet as well as maintained it. But the emergence of

Figure 10   The Belfast truss roof of a surviving building at the former Handley Page factory at Chadderton, Manchester (Photograph: Roger J C Thomas)

Germany as a serious maritime rival and threat at the end of the nineteenth century led not just to an accelerated naval arms race, but also to a radical restructuring of the fleet. Admiral Sir John Fisher's reforms refocused the fleet away from the Empire and towards the North Sea. Amongst the most visible reforms were the modernising of the Channel Fleet based at Dover, and the sending of the new Dreadnoughts to the Chatham division of the Home Fleet. Rosyth Dockyard was created to serve the Home Fleet based at Invergordon, with Scapa Flow earmarked as the main northern base in the event of war with Germany. Less visible was the continuing modernisation of the main dockyards at Chatham, Portsmouth and Devonport. Dry docks and slips were augmented and enlarged, extended berthing facilities provided, and new naval barracks and workshops built. The latter particularly included electrical, torpedo and gun-mounting factories. Oil fuel depots gradually replaced the coaling stations.

In the aftermath of World War I, the Washington Naval Treaty of 1921 limited the size of the world's largest navies and effectively ended Britain's two-power standard. In the inter-war period, little money was available for modernising the fleet facilities; major dockyard construction was limited to Singapore. One of the very few inter-war dockyard buildings of any size was the still-standing No 4 Boathouse at Portsmouth. The experience of World War II demonstrated that Britain no longer had the capacity to fight and win a major conflict unaided, a fact recognised by the joining of NATO in 1949.

Since 1945, the Royal Navy has undergone a very steep numerical decline, although with its nuclear role its destructive power is greater than ever. Overseas bases have been closed, with the exception of residual facilities at Gibraltar. Victualling yards, naval hospitals and barracks have been shut down. Faslane has been created as a base for Trident submarines and Rosyth and most of Devonport are run by private contractors. In the 1960s, nuclear refitting facilities, subsequently demolished, were built at Chatham; more recently, Devonport has been modernised to maintain nuclear submarines. The end of warship building in the royal dockyards in the mid-1960s reduced the need for workshops and heralded the shutting of foundries, the demolition of building slips as well as the closure of the dockyard technical schools and later the Royal Naval Engineering College at Manadon. However, at Portsmouth naval base, a commercial warship builder is currently establishing shipbuilding facilities in the northern area of the base, in part using existing facilities. Chatham Historic Dockyard, in the shape of 7 Slip and No 1 Smithery and its contents, probably now has the best surviving evidence for twentieth-century warship building in the royal dockyards. Apart from nuclear facilities and the modernising of naval barracks, the only large-scale post-war

Figure 11   Grade II* listed hangars at Calshot, Hampshire, built in 1917 for housing Felixstowe F5 flying boats. These are now used as part of an Outward Bound School (Photograph: Mike Williams © English Heritage)

dockyard construction projects have been a series of architecturally distinguished workshops at Portsmouth in the 1970s and the contemporary Frigate Refit Complex at Devonport. More recently, the North Camber Project at Portsmouth has provided extensive new berthing spaces.

The naval dockyards of the pre-twentieth century have been the subject of characterisation and assessment projects by David Evans and Wessex Archaeology (eg Wessex Archaeology 2000a and b). Although these studies have included reference to the yards' later use, their focus has been on earlier periods.

[OBJECTIVES: A3, A10–11, A13, B1, B12, D2–4, E2, E4–5, F]

Other naval installations

Naval installations were not confined to the dockyards or their immediate surroundings. In the 1890s, the Admiralty had instituted a series of three different types of war signal stations around the coasts. These were designed for flag communication between naval and merchant ships and the shore. At defended ports, they had the added responsibility of identifying

Figure 12   Harperley prisoner of war camp, County Durham. The site is now a Scheduled Monument (Photograph: Margaret Nieke)

warships. Signal stations remain at Dover and Falmouth, amongst other places. In World War II, a string of radio intercept stations was established to intercept U-boat radio traffic, to obtain accurate coded texts of transmissions, and to obtain bearings on the sources. In all, there were some 51 stations on both sides of the Atlantic. The Admiralty also operated a number of its own shore-based radar stations; Dover Castle and Beacon Hill, Harwich contain associated structures.

Early experiments in naval aviation led the Admiralty in 1912 to propose a chain of seaplane bases along the coast from Scapa Flow to Pembroke; Calshot is an early example. These were joined later by naval air stations for carrier and land-based aircraft.

The two world wars caused an immediate and enormous expansion of naval installations. Many of these were to do with training. Hutted camps were frequently established around requisitioned country houses and generally have left little trace. In World War II, boarding schools, such as Roedean and Lancing College, were taken over, as were other facilities such as the King Alfred Baths at nearby Hove. When Dartmouth was bombed, the college moved to Eaton Hall, Cheshire (now demolished). Research facilities also expanded rapidly. For example, existing ship-model tanks at Haslar were supplemented and later largely superseded by new research facilities at Teddington. Naval

radar test facilities were established alongside Fort Cumberland and at Eastney Fort East at Southsea.

In the late 1930s the Admiralty began to put its more vulnerable operational headquarters underground as a protection against bombers. Where possible it used existing spaces. Good examples, now open to visitors, are Admiral Ramsay's headquarters below Dover Castle and the Atlantic convoy operations headquarters in Liverpool. Further north, Pitreavie Castle became a Maritime Combined Headquarters. Other underground naval headquarters exist at Chatham (HMS *Wildfire*), Newhaven, Portsmouth and Devonport. Those at the main naval bases were later adapted for Cold War use. Northwood, for example, became the headquarters of the major NATO Command CINCCHAN (Commander in Chief Channel). In 1939, following experience gained in World War I, extensive controls were in place for convoy operations. The Ocean Convoy System, based on the Clyde and in Liverpool, controlled the east–west route across the Atlantic and the north–south route to Freetown (Sierra Leone). In general, existing buildings appear to have been requisitioned to provide many of the shore facilities for these convoy systems. Closely associated were the mine barrages. Those down the east coast and across the Straits of Dover were the most extensive, but blockade minefields and trap minefields were also laid. All these required shore facilities associated with mine storage, the operation of boom defences, the laying and sweeping of mines and the housing of naval personnel. In September 1939 there were 131 operational minesweepers and converted trawlers based at 25 ports and dockyards around the UK. These numbers were to rise in World War II.

Also relevant here is the importance of coastal traffic around the British Isles in both world wars. The transportation of goods by coastal shipping overcame limitations and overloading of the road and rail systems. For example, the Southern Railway imported coal by sea from south Wales to ports in Devon and Somerset to minimise rail congestion in the Severn Tunnel.

[OBJECTIVES: A3, A10–11, A13, B1, B12, D2–4, E2, E4–5, F]

Camps

Camps are one of the least studied categories of site, perhaps because typologically they form one of the most diverse, and one of the largest in number. To date it has proved impossible to find any overall figures for the numbers of camps constructed in either World War I or II, or any convenient list of their locations. For World War II, from the raising of the militia in April 1939 (which demanded 33 new hutted camps), through the return of the British Expeditionary Force from Dunkirk, to the build up of forces for the D-Day landings (including overseas troops, of which the Americans were the principal numeric component), many hundreds of hutted and tented camps were constructed. Required additionally were camps for anti-aircraft, searchlight, and coast batteries, for balloon squadrons, for prisoners of war (Hellen 1999), conscientious objectors, camps for refugees and displaced persons, for internees, and for the Women's Land Army and other labour organisations or war factory workers. The number of camps in Britain, loosely defined, potentially runs into the thousands.

In addition to these purpose-built camps, there was much extension to existing barrack accommodation (see Douet 1998 for a discussion of barrack accommodation to 1914, though noting that the term 'camps' can specifically exclude permanent cantonments and barracks, cf *Military Engineering* V, 1934). Many country and smaller houses, hotels and even castles (Lloyd 2001) were also requisitioned to serve as army headquarters, with camps (often tented) built in their grounds. Troops would also be billeted with private householders or accommodated in barns.

A starting point for the study of camps is the model plans, specifications and *Barrack Synopses* issued by the service ministries, often disseminated as printed manuals. Closely associated are designs for buildings, ranging from prefabricated hutting to 'camp structures' for improvisation in the field: latrines, ablutions, water points, straw stores for

palliasses, incinerators, cookhouses, rations' stores, armouries and so on. These sources obviously embody ideals; reality was often different and departures from the formal models are of great interest in themselves. Often recorded (or lamented) in War Diaries, the actuality of troop accommodation can also be recovered from aerial photographs, oral history, large-scale mapping and archaeology. An English Heritage study of PoW camps (whose documentation is good), has assessed what survives today (Thomas 2003) and, *inter alia*, has confirmed the insights that wall art can give to camp life.

[OBJECTIVES: A1–2, B1–2, B8. B13–14, D, E2, E4–5, F]

Hospitals

A study of English hospitals, 1660–1948, has recently been completed (Richardson 1998). This includes a survey of military hospitals placing twentieth-century examples within their broader temporal and thematic context. The report describes the late nineteenth- and early twentieth-century hospitals, which embodied the main principles of pavilion planning. The Herbert Hospital at Woolwich (completed in 1865) was to this plan, as was the Cambridge Military Hospital at Aldershot. The Queen Alexandra Military Hospital on Millbank, London (1903–05) and another of that name on Portsdown Hill (1904–07) were the last general military hospitals to be built on the pavilion system. Later general military hospitals were either hutted or occupied converted premises.

Between 1860–1914 a system of regimental and barracks hospitals was utilised. However the regimental hospital system was abolished in stages between 1870–73 as part of the general reorganisation of the army, and in their place numerous small establishments, more sick bays than hospitals, were erected to serve regimental districts throughout the country.

In 1907, in preparation for war, the Royal Army Medical Corps set up 23 territorial hospitals in existing buildings. Providing some 12,000 beds these would augment existing beds in military (9000) and voluntary (10,000) hospitals available in wartime. It was recognised that many of the temporary establishments would have to be extended by hutted accommodation, for which the War Office drew up model plans. When war broke out, hutted hospital accommodation was provided with remarkable speed and economy in the grounds of asylums, colleges, hospitals and private houses. The model plans dictated that most had parallel rows of pavilion wards while others adopted a semi-circular arrangement. Most buildings had timber frames but with a variety of facing materials including corrugated iron, asbestos sheeting and brick. In 1915 several experimental open-air wards were used with the aim of extending open-air treatment.

After World War I the number of military hospitals in England was reduced and in some

Figure 13   Entrance to the military hospital beneath Dover Castle. The hospital saw much use during the later years of World War II and is now a popular visitor attraction (Photograph: John Schofield)

stations the hospital facilities of all three services were concentrated in one establishment. By 1935, when the War Office began to prepare once more for war, there were only 3000 equipped military hospital beds in the country. Under the Emergency Medical Service (EMS), wartime hospital accommodation was provided for the military and for civilians alike, given the likely need to treat air-raid casualties. Some 86,000 civilians were seriously injured in World War II through enemy bombing and long-range bombardment. Hospital trains were also equipped and stabled at key points, to evacuate civilian air-raid casualties, but later in the war military trains were used to bring the wounded back from the invasion ports to hospitals in Britain.

On its creation in 1918, the RAF had no general hospitals of its own, though RAF stations were provided with sick quarters, usually accommodated in Nissen huts or other temporary structures. However, from 1918, purpose-built RAF hospitals were built. Several hutted RAF hospitals were constructed during World War II, one of the largest at RAF Cosford. The social history and memory of a late nineteenth- and early twentieth-century hospital at Netley, Hampshire is captured in Philip Hoare's book *Spike Island* (2001).

[OBJECTIVES: B11, D3–4, E2, E4–5, F]

Intelligence

The built infrastructure of intelligence has not so far featured in our assessments, but the whereabouts of successive headquarters and many other facilities are readily gleaned from published histories. British intelligence in its modern form is a creation of the twentieth century – an era of multiplying effort, professionalism, formality and technological finesse. 'Professional' intelligence in Britain began around the start of World War I, and by the 1940s a complex structure was in place, with sizeable directorates active in the three service ministries, and the Government Code and Cipher School (GCCS) established at its famous refuge of Bletchley Park. Today, under the unified Ministry of Defence, military intelligence is overseen by the Defence Intelligence Staff, while civil branches exist in the Security Service and Secret Intelligence Service. The GCCS meanwhile has become the Government Communications Headquarters (GCHQ).

[OBJECTIVES: A3, A7, A9, A13, B1, B6, B12, D3, E2, E4–5, F]

Communications

The value of archaeological study at broadcasting and communications sites is beginning to be realised (eg

Martin 2002) though it remains an area where much research is needed, covering themes such as wireless (or radio) communications, telegraphy, telephone and wireless, satellite, and the increased importance of computer-based communications through the Cold War and post-Cold War periods. The use of fixed (landline) telephone networks and in particular that provided by the GPO, its predecessor, and BT were (and still are) crucial to the British defence effort. Much effort was invested in improving the resilience of the GPO before and during World War II and the Cold War period. In the latter stages of the Cold War the new operators (Mercury, Cellnet and Vodafone) played a small part. Infrastructure included underground and hardened facilities such as trunk exchanges, international exchanges or repeater stations. There was diversity of routing and, later, the use of microwave was introduced as an alternative to landlines. Submarine cables for international communications were hardened at their landfall points. The armed forces also ran their own telephone and telegraph networks (for example the Defence Telegraph Network) using rented civilian bearers, and established their own trunk exchanges and tape relay centres. Many of these were in protected accommodation and some of the sites survive.

The use of radio communications to provide both tactical and strategic communications by voice and telegraph is a major subject. It covers such areas as ship to shore, submarine broadcast, ground to air, air to air and army tactical communications within formations and units often down to individual fighting vehicles. In addition to commercial services run, for example, by the Post Office or Cable & Wireless, the armed forces each ran their own global networks with major transmitter stations, receiver sites and tape relay centres in the UK. For example, the Commonwealth Army Communications Network or COMCAM had Boddington (a hardened tape relay centre in Gloucestershire) with radio sites at Bampton and Droitwich.

[OBJECTIVES: A3, A7, A9, A13, B1, B6, B12, D3, E2, E4–5, F]

Theme 4: Operations

Shipping

A considerable amount of twentieth-century naval equipment survives in specialist public and private collections. For ships, the *Core Collection* of historic ships (accessible via www.nhsc.org.uk) includes a number of twentieth-century vessels of very great naval and mercantile significance. Among these are *Excelsior* (built at Lowestoft, sailing trawler, 1931), *Robin* (built at Blackwall, tramp steamer, 1890),

Figure 14   SMS Coln, *one of seven wrecks from the World War I German High Seas Fleet scuttled in Scapa Flow (Orkney) and scheduled under the Ancient Monuments and Archaeological Areas Act 1979. Top: multi-beam sonar plot of the wreck in 2001 (courtesy of ScapaMAP). Bottom: contemporary photograph, reversed for comparison (courtesy of Charles Tait)*

Turbinia (built at Newcastle, Charles Parson's turbine-powered vessel, 1893), HMS *Alliance* (built at Barrow-in-Furness, A-class submarine, 1945), HMS *Belfast* (built at Belfast, 6-inch gun cruiser, 1938), HMS *Caroline* (built at Birkenhead, Light cruiser, sole surviving veteran of the Battle of Jutland, 1914), CMB4 (built at Hampton, shallow-draught motor boat armed with torpedo, 1916), *Holland 1* [built at Barrow-in-Furness, Royal Navy's first submarine, 1901], HSL 102 (built at Hythe, High Speed Launch, 1936), *Landfall* (built at Hepburn, tank landing craft, 1944), *Medusa* (built at Poole, HD Motor Launch, 1942), *Minerva* (built at Belfast, shallow draught monitor, 1915), MTB 102 (built at Portsmouth, prototype of Vosper-built MTBs, Dunkirk veteran, 1937) and T3 (?built at Richborough, one of 45 seaplane lighters, towed at speed to allow launch of Sopwith Camel, c1911). Two other naval vessels, not on the *Core Collection,* but of considerable importance for their naval associations are HMS *Cavalier* (built at Cowes, last World War II destroyer, 1944. On the *Designated Vessels* list of the National Register of Historic Vessels) and HMS *Ocelot* (Chatham, last submarine to be built at Chatham for the Royal Navy, 1962). Both of these are on display in the heart of Chatham Historic Dockyard.

On 30 September 2002 eleven military maritime graves were designated as Controlled Sites under The Protection of Military Remains Act (1986), including: HMS *Dasher*; HMS *Exmouth*; and HMS *Formidable*. Further vessels (lying outwith UK territorial waters) were designated as Protected Places.

Recent warships have also been protected as Designated Historic Wrecks under The Protection of Wrecks Act (1973) such as: the early aircraft carrier HMS *Campania*, and the submarine HMS A1. A further submarine, the *Holland 5*, is in the process of being designated.

The remains of many thousands of other vessels, a significant proportion of which will be military in function, exist on the seabed around the UK and other countries, and in international waters. An impression of the former category can be gained from preliminary survey work on hulks preserved on the foreshore of the Tamar (Martin Read, pers comm). Of approximately 50 vessels recorded, six were of military origin including landing craft, ammunition barges and the steam pinnace from the World War II French battleship *Paris*. A similar survey of the Kingsbridge estuary revealed landing craft and other vessels used as the moorings for barrage balloons on the Dart.

[OBJECTIVES: A11, B3, C2, D2–3, E4–5, F]

Preparations for D-Day

Between midnight on 6 June (D-Day) and 30 June 1944, over 850,000 men had been landed on the invasion beachheads of Normandy, together with nearly 150,000 vehicles and 570,000 tons of supplies. Assembled in camps and transit areas over the preceding months, this force was dispatched from a string of sites along Britain's coastline between East Anglia and south Wales (Dobinson 1996d, 2). Those sites in England involved in this embarkation have now been studied with a view to identifying which survive and affording them appropriate protection (Dobinson op cit; Schofield 2001).

Of these D-Day preparatory sites, there are four principal classes of monument which represent key aspects of the operation. Surviving as impressive monumental remains they also symbolise the scale and international significance of the events of June 1944. The four classes can be characterised in the following terms:

Mulberry harbour construction sites

The construction of the two artificial 'Mulberry' harbours, built in sections (different component parts generally at separate sites) and towed across the channel for disembarkation of troops and landing of supplies, were in Churchill's words, 'a principal part of the great plan', and were decisive in the first days of the invasion. Although one harbour failed, the remaining structure – at Arromanches – was significant in providing the tactical advantage of surprise, and the logistical advantage of not having to land on a defended shore and at the mercy of the weather. Some components of the harbours were surplus to requirements and remained in the UK; some sank on route, or were 'beached' for other reasons. Many sites were involved in this construction process, stretching at least from Southampton, via south coast ports and London, to the North East. These construction sites were located either in largely unmodified dry docks or slipways, or in excavated basins or beaches. Much use was made of existing facilities, with security, camouflage and deception being of vital importance. It is the beach construction sites, however, that retain most evidence for this construction task, comprising construction platforms, slipways and winch house foundations. Some of the components built for the harbours also survive, mostly at sea, having sunk while on tow, but occasionally on land, as with the line of 36 'Beetles' at Dibden, Hampshire. Sunken Mulberry debris has been noted by recreational divers at various locations off the south coast (see refs in Schofield 2001). Of course, parts of the Mulberry harbour at Arromanches survive *in situ*.

[OBJECTIVES: A1, A13, B1, B3, C2–4, D3, E2, E4–5, F]

Maintenance and repair areas

The maintenance and repair areas, and harbours used for landing craft and landing ship construction, were

Figure 15   D-Day embarkation slipways at Torquay Harbour (© English Heritage, AA012296)

essential to developing and retaining a fleet capable of delivering Churchill's 'great plan'. With so many vessels involved (landing craft and landing ships principally, but 46 different types of vessel in all), maintenance was a significant task. Contemporary descriptions talk of unprecedented levels of maritime activity, with every port, harbour and boatyard being involved. In addition the maintenance areas extended to many beaches, where slipways and gridirons were built. Even the streets of coastal towns and villages were used for repairs.

The purpose-built gridirons were used for maintenance, and took the form of a series of parallel concrete rails running down a slight gradient into the water, allowing a boat to be floated on at high tide, and repaired at low tide; some were supplied with a winch mechanism for pulling vessels onto the grid, and timber and steel mooring points ('dolphins') for securing them when afloat. Repair areas in the form of slipways, with a metal rail, winch mechanisms and dolphins, are known to have been used for landing ship repairs. However, much of the repair and maintenance activity was conducted on an *ad*

hoc arrangement and leaves little trace: for example, landing craft (assault) – LCAs – were small vessels constructed and repaired mainly in back streets and on improvised hards at the water's edge.

[OBJECTIVES: A1, A13, B1, C2–4, D3, E2, E4–5, F]

Embarkation sites

Embarkation sites had to be well-designed and well-built if embarkation was to be a rapid and efficient exercise. Geographically the sites had to have access to hinterlands in which large numbers of troops and supplies could be concealed from enemy reconnaissance, yet which had the road and rail networks to allow their easy movement at the time of departure. This part of the operation was planned well in advance, with most embarkation hards built in the period between October 1942 and the spring of 1943. In all 68 embarkation sites are documented in public records (Dobinson 1996d), representing those built specifically to serve general cross-Channel operations from 1942 onwards, and the extension to that group built to serve Operation *Neptune*. The list is complete in both these respects. However, as photographs show, embarkation also took place at other sites not built for the purpose.

Embarkation sites were either modified docks, quays or harbours (such as Southampton docks) or were constructed specifically for the purpose. Two main types of loading facility were used: LCT hards for 'landing craft, troops' and LST hards for 'landing ship, tanks'. Although LST hards were the most numerous, the two types were broadly similar. Each had: a concrete apron (solid concrete above high water, and flexible concrete matting below), and a series of dolphins; hutting for offices, workshops and stores; fuelling facilities; electric lighting, and roads and transit areas (see Dobinson 1996d, 14–18 for details). Survival tends to be confined to those hards built specifically for the purpose (those in existing docks having been redeveloped in the post-war period).

Logistics, camps and supplies

Finally there is PLUTO, the Pipeline Under the Ocean, the logistics system that underpinned the entire operation, and camps established and occupied in the months prior to embarkation. PLUTO (and SOLO – the Pipeline Under the Solent, a crucial link in the network) was established to provide fuel for the invasion force, and took the form of a complex and extensive system of pipelines and terminals, with pumping stations at Dungeness (Kent) and Sandown and Shanklin on the Isle of Wight (see Searle 1995 for details). Although the pipelines were cleared from the seabed after the War, short sections do survive, particularly off the Isle of Wight. A pipeline valve survives at the Hamble oil refinery in Hampshire, while at the SOLO terminal at Thorness Bay on the Isle of Wight, shore-end pipe connections are exposed at low tide (Searle 1995, 84).

Little has been recorded to date of the logistics systems that underpinned the operation, beyond the work on PLUTO and the port and maintenance facilities for craft, both outlined above. In addition to the pipelines, these logistics systems included depots, stores, parks, dumps, fuel storage sites, rail heads, sidings and rail interconnects. Production and support facilities were also important, notably: laundries, bakeries and printers for cartographic and hydrographic charts.

Unsurprisingly, little will survive of the many camps occupied prior to embarkation. What do survive however are the sources that reveal their locations and the effect of this encampment on the contemporary landscape. Aerial photographs by the United States Army Air Force (March 1944) of the Truro area in Cornwall show the hundreds of bell tents occupied by US troops concentrated along arterial routes. Contemporary maps and plans, and ground photographs, show the overall layout, and the character of these sites (see various photos in Doughty 1994). Finally, some hutted camps and hospitals do survive, though often now as developed sites: industrial estates, modern hospitals and garages.

Work sponsored by English Heritage has also taken us further towards understanding how one of the most well-known aspects of *Overlord* appeared on the ground. This was Operation *Fortitude* – the cover plan for the *Neptune* landings and subsequent cross-channel activity – which in addition to a wide range of strategic deception techniques employed lighting decoys and dummy displays at sites around the south coast (Dobinson 2000b).

[OBJECTIVES: A1, A2, A13, B1, B5, B11, C2–4, D3, E2, E4–5, F]

Air defence

As part of the Monuments Protection Programme, a major study of the related monument classes within this category and others was undertaken in 1995–99 (Dobinson 1995, 1996a, 1996b, 1999a, 2000b, 2001; Schofield 2002a). For English sites the study was in several parts, divided according to monument class, and into two distinct stages of work:

Stage 1: characterisation based on archive sources
Stage 2: evaluation based on modern aerial photographs.

A third stage, involving further evaluation and an assessment of future management needs was based on subsequent field visits.

For Scotland, Wales and Northern Ireland work has been completed to Stage 1, as a follow-on from the MPP study (Redfern 1998a–d). In addition, Historic Scotland and RCAHMS have completed a field study of the defences of Scotland, and are adding to this with information from German and British contemporary aerial photographs. Some additional work has also been undertaken in Northern Ireland.

The Stage 1 work was an extensive consultation of documentary sources bearing upon the location, form and context of military sites, chiefly from the holdings of the National Archives (PRO). Too broad for easy summary, this material ranged from unit records, through headquarters' papers to governmental archives at the higher end. Information on location largely comprised six-figure references on the wartime Cassini grid; many thousands of references – for air defence sites, anti-invasion works, coast artillery batteries and others – were extracted and duly converted to modern NGRs. Though this substantial archive had been routinely used by historians since the bulk release of World War II papers in 1972, it had never been used for this purpose on so comprehensive a scale. It produced something approaching a definitive list.

Follow-up work at Stage 2 involved checking the

Figure 16   The cropmark and slight earthwork of a 40 mm Bofors gun pit at RAF Staxton Wold, East Riding (Photograph: Roger J C Thomas)

Figure 17   Curved asbestos huts surviving at a World War II Diver site on the Essex coast. In this case the gunsite has been removed and only the domestic accommodation remains (Photograph: John Schofield)

grid reference for each site on recent (and where available immediately post-war) aerial photographs to determine modern survival. Sites were graded according to the completeness of what remains compared to what type-drawings suggest was present originally. The results tell us a great deal about the fate and state of a range of wartime monuments.

Anti-aircraft artillery

Britain's first anti-aircraft guns were emplaced before the First World War, at which time they defended only military targets. Although not quantified as part of English Heritage's work on anti-aircraft artillery, a total of 376 locations in England have been identified in archives dating from this period; very little may survive, however, following the wholesale run-down of Britain's defences after the Armistice in 1918. The great majority of gunsites date from World War II, when a variety of weapons was organised to meet the bombing of major towns and cities, manufacturing areas, ports and airfields. Three main types of site existed in this period:

- heavy anti-aircraft (HAA) – big guns to engage high flying bombers
- light anti-aircraft (LAA) – defence against lower flying aircraft
- 'Z' batteries (ZAA) – relied on the shotgun effect, using density of rocket fire against both low and higher flying targets

MPP's assessment has provided grid references for 3188 sites built in England (excluding 272 light anti-aircraft positions relating to airfield defence), in addition to the 376 general locations (eg on earlier forts) where World War I anti-aircraft gunsites were located. Most of the World War II sites were abandoned in 1946, although in England 192 heavy anti-aircraft batteries continued in use after the war, forming part of the Nucleus Force; these guns were finally stood down in 1955 when defence against a new generation of high flying Soviet bombers was better achieved with guided missiles. A further 458 sites are documented for Scotland, 166 for Wales, and 71 for Northern Ireland (Redfern 1998a, 3).

This is a diverse and significant group of monuments (Dobinson 1996b, 2001). The diversity in form and layout was determined mainly by the number and type of weapons the gunsites contained. The weapons dictated the facilities needed within the operational site, and the number of people needed to operate guns and instruments (such as radar). In turn personnel levels determined the nature and scale of domestic accommodation (the armed forces having well-laid-down Synopsis Scales for Accommodation). Significant developments in anti-aircraft artillery can also be seen in the physical remains, as progress and change, social and technical, are reflected in the equipment and layout

of gunsites. The greatest technical advance during World War II was the development of radar for target location and fire control; and radar was introduced to some gunsites as early as September 1939. Also the employment of women on operational gunsites was a milestone in Britain's evolution as a military power, and this is reflected in the arrangement of facilities on some domestic sites.

Of the 981 heavy anti-aircraft sites in England for which grid references were established through the study of primary sources, only 57 (5.8%) are revealed as now being either complete or near complete (see Dobinson, 2001 for more detail; and Anderton and Schofield 1999). A further 119 (12.1%) have partial remains, while for 14 examples photographs and map-based work was inconclusive. In total 81.7% of heavy anti-aircraft gunsites are recorded as having been removed, mostly the result of post-war development (40% of removed sites) and agriculture (43%). In Northern Ireland at least ten sites survive in good/ fair condition. The sites recorded as complete or near complete in England include two of the first eighteen mixed batteries authorised in one AA Corps. Of the 192 sites selected for post-war use as the Nucleus Force, 30 survive as complete or near complete examples. This figure (15.6%) contrasts with the 5.8% of total heavy anti-aircraft sites surviving in this condition. Not surprisingly, therefore, retention as part of the Cold War Nucleus Force has aided preservation.

Very few light anti-aircraft sites survive in England. Of the 966 recorded sites (excluding airfield defences – the subject of a separate study) only three (0.3%) are recorded as complete or near complete, 42 (4.3%) as having partial remains, and there are 28 sites for which this method of checking proved inconclusive. In total 92.4% are recorded as destroyed.

From an indicative list of 51 'Z' batteries documented for England, all were recorded as destroyed by this survey. One surviving example has subsequently been found, by chance, by a separate field survey (Schofield *et al* 2001). The most complete site yet discovered is on Flotta in Orkney.

Diver sites were improvised or constructed to meet the ever-changing reality of the flying bomb. Of the 1190 sites identified in documentary sources, nine (0.8%) are complete or near complete, 72 (6.1%) have partial remains, and for 32 aerial photographs and maps proved inconclusive. Not surprisingly given the mobility of the *Diver* campaign, heavy anti-aircraft batteries survive better than those for light weapons, for which nothing survives complete or near complete. Also, the greater permanence of sites along the east coast is reflected in higher survival rates. For heavy anti-aircraft batteries, 6% of the 187 sites in the Coastal Gun Belt (Kent, Sussex) survive, contrasting with 40% of the 81 sites on the *Diver* Strip (Essex, Suffolk). In all 90.7% of *Diver* sites are recorded as destroyed, 60.5% lost as a result of post-war farming practices.

As an adjunct to anti-aircraft artillery, searchlight

emplacements and barrage balloon sites are comparatively well understood. Lists of these sites do exist in public records, and some of the searchlight emplacements have been documented (Dobinson 1999c). However, few surviving examples have been recorded and no systematic attempt made to identify them for conservation purposes.

For all of these types of anti-aircraft site it is anticipated that survival rates will be comparable in Scotland and Wales to those for England and Northern Ireland. If anything the rates might even be higher, given the relative intensity of post-war development.

[OBJECTIVES: A5, A12–13, B1–2, B5, C2–4, D1–3, E2, E4–5, F]

Acoustic detection and World War II radar stations

Early-warning was fundamental to Britain's strategic air defence from the First World War (Dobinson 1999a) when visual spotting gave way to sound mirrors. The distribution of the earliest acoustic sites is imperfectly known, but an experimental 1920s group on the Kent coast has received much study. It was these devices which were eventually supplanted by radar.

The principles behind radar were widely recognised by the 1930s – being that an electromagnetic pulse reflected from an object betrays the object's position to a receiver – and in Britain this was developed into a practical adjunct to air defence. Following experimental work at Orford Ness and Bawdsey in Suffolk, ground radar for surveillance and early warning developed during World War II into six main areas, each distinctive in plan-form and in the buildings and structures they contain:

Chain Home – the original 'chain' of radar stations, which developed and expanded through the war years. This was designed for raid reporting, passing information to a central operations room which in turn directed fighters to intercept enemy aircraft
Chain Home Low – developed to fill gaps in low-level cover left by the original technology
Ground Controlled Interception – an adaptation of radar during the Blitz of 1940–41 by which night fighters were controlled directly, rather than via a central control room
Coast Defence/Chain Home Low – also added in 1941 as a low level cover coastal radar designed to detect aircraft and surface shipping
Chain Home Extra Low – which involved the conversion of many existing stations to take new and more powerful equipment
Fighter Direction radar – introduced in 1943 to aid Fighter Command in their offensive sweeps over occupied Europe.

In all, primary sources confirm that 200 locations in England were occupied by 242 separate radar

reporting and control functions during World War II, many of which continued in use into the Cold War period (see below). A further 71 sites are documented for Scotland, 26 for Wales and 12 for Northern Ireland (Redfern 1998a, 3).

Follow-up work in England using aerial photographs has shown that some 105 of the 242 World War II radar stations in England, as recorded on the most recent aerial photographs, have been removed, the great majority by human agency and on land subsequently used for arable cultivation (see Schofield 2002a, 271–3 for more detail). However, this contrasts with the figures for other wartime remains (bombing decoys and anti-aircraft gunsites for example), with significantly more of these radar sites surviving in some form at the time that the photographs were taken. For many this is due to the continued use of radar stations through the Cold War period (though with significant refitting and alteration); a few remain in use today.

[OBJECTIVES: A13, B5, C2–4, D1–3, E2, E4–5, F]

Bombing decoys

Britain's decoy programme began in September 1939 and developed into a complex and diverse deception strategy, using three main methods (Dobinson 1995, 2000b):

–   Dummy structures and features ('K' sites, dummy factories and wireless installations)
–   Lighting ('Q' sites and 'QL' sites, and their derivatives in *Fortitude* etc)
–   Fire ('QF' sites and 'Starfish')

In all, some 839 decoys are recorded for England in primary records, built on 602 sites (some sites containing decoys of more than one type). This makes up the greater proportion of the decoys recorded for the United Kingdom, 74 of which are in Scotland, 54 in Wales and nine in Northern Ireland (Redfern 1998a, 3).

The programme represented a large investment of time and resources. Apart from construction costs, several thousand men were employed in operating decoys, 695 of which were simultaneously active at the height of the campaign in November 1942. In common with much wartime technology, the decoy programme began from almost nothing yet rapidly achieved a high degree of technical refinement. Furthermore, decoy sites are closely tied to the wartime fortunes of the targets they served. The decoys were often successful, drawing many attacks otherwise destined for towns, cities and aerodromes. Examples survive on the north bank of the Humber, east of Hull, and on Blackdown on the Mendips (Schofield *et al* 2001).

Follow-up work in England has shown that at least 505 of all bombing decoys, as recorded on the most recent aerial photographs available through the

Figure 18 (above)   Documents combined with the form of the building at this site on the Cornish coast reveal it to be the remains of a Coast Defence / Chain Home Low and Chain Home Extra Low radar station (Photograph: John Schofield)

Figure 19   All that remains of this former bombing decoy at Menthorpe, Yorkshire is the brick-built control building (Photograph: Roger J C Thomas)

National Monument Record, have been removed, again mostly by human agency and on land subsequently used for arable cultivation (see Dobinson 2000b and Schofield 2002a, 276–7 for more detail). Although 189 of the decoys survived in some form at the time the photographs were taken, for the majority this amounted to nothing more than the control building. In general it is the night-time decoys that have survived best, not surprisingly given that it was these that were equipped with built structures and a more substantial form of decoy. (For aerial photographs of decoy sites, both in 1946 and today, see Dobinson 2000b.)

[OBJECTIVES: A13, B5, C2–4, D1–3, E2, E4–5, F]

Civil defence

Documentary research (Dobinson 1999d) has established the chronological and typological framework for civil defence provision in World War II, while a recently completed project in Southampton, based on local archives to determine where shelters were constructed, followed by fieldwork, suggests that between 6–10% of the characteristic Anderson shelters survive, the majority in poor condition (Lacey 2002). Beyond these projects however, and other site-specific research initiated through the planning system, or to support listing recommendations, little is known of what survives.

For the Cold War, the principal types of central government emergency headquarters were identified in English Heritage's survey (Cocroft and Thomas 2003). In this study, examples of local authority emergency accommodation at county and district level were recorded, but no definitive list of all the structures of this type was produced. Similarly, various post-war civil defence structures were recorded; anecdotal evidence suggests more examples of such structures will emerge through research at local record offices. Documentary evidence also indicates that some households did build private nuclear shelters, but few examples are known.

[OBJECTIVES: A3, A13, B1, B10, B12, C2–4, D3, E2, E4–5, F]

Anti-invasion defences

Between 1994 and 2002 the Defence of Britain Project collected data on twentieth-century defences in Britain, with a focus since September 1998 on anti-invasion defences, building on earlier work by Henry Wills and others (see www.britarch.ac.uk/projects/dob/review/index.html for the project review). At the same time English Heritage sponsored the documentary study reported in Dobinson 1999c, and have included anti-invasion defences in the National Mapping Programme (notably in coastal East Anglia). A model field survey has also been completed in Essex (Nash 2002). Although the bulk of records relate to World War II, there are some records of surviving World War I pillboxes on the database. While Britain was not fortified during World War I as it was to be in World War II, half a million men were nevertheless held in reserve, and lines of pillboxes, with accompanying earthworks, are known to have been constructed in Yorkshire, East Anglia and Kent against a possible invasion threat. Coast batteries were also built or recommissioned. The bulk of activity, and thus surviving remains, dates from the early part of World War II, however, with the invasion threat lasting from the period after Dunkirk (June 1940) until 1940/41, even though the perceived threat lasted to 1943 when the Joint Intelligence Committee ruled out this possibility. The building of anti-invasion defence works was therefore carried out very quickly, and in accordance with defence planning that itself was altering radically and rapidly through daily adaptation to military necessity, the build up of military resources, and the ideas and policy of newly appointed commanders.

In broad terms, the defence landscape has to be viewed through the military areas, districts, sub-areas, sub-districts, and sectors by which it was controlled. Each military ground area was subject to a detailed defence scheme, and each military unit itself maintained a defence scheme for the area it was to defend, extending from Army General Headquarters (GHQ) home forces, through the commands down to an infantry platoon or section. In addition, each fortification had a scheme for its defence. In this way, all of Britain fell within a defence scheme, and every structure or work was categorised, referenced, mapped, and its precise role detailed in writing. Many of these defence schemes and plans survive amongst public records.

Ground defences evolved to embrace a variety of types, dependent on a combination of function, form and location. Defence of the coast was of prime importance, to stop the enemy landing at all, or to pin him to a coastal strip. Stop line defences were principally anti-tank check lines, designed to confine enemy columns to particular areas of the hinterland: stop lines usually followed topographic features, generally river valleys or canals – they were set up within military commands or divisional areas, and often divided one element of the field army from another. The GHQ Line was the main stop line, having significance beyond the considerations of army command or military area – its principal purpose was to protect London, the Midlands, and the industrial north from attack following landings on the south or east coasts. The major feature of the stop line was the continuous anti-tank obstacle, often a waterway, that might be artificially strengthened, or a machine-dug anti-tank ditch – gaps in this earthwork would be

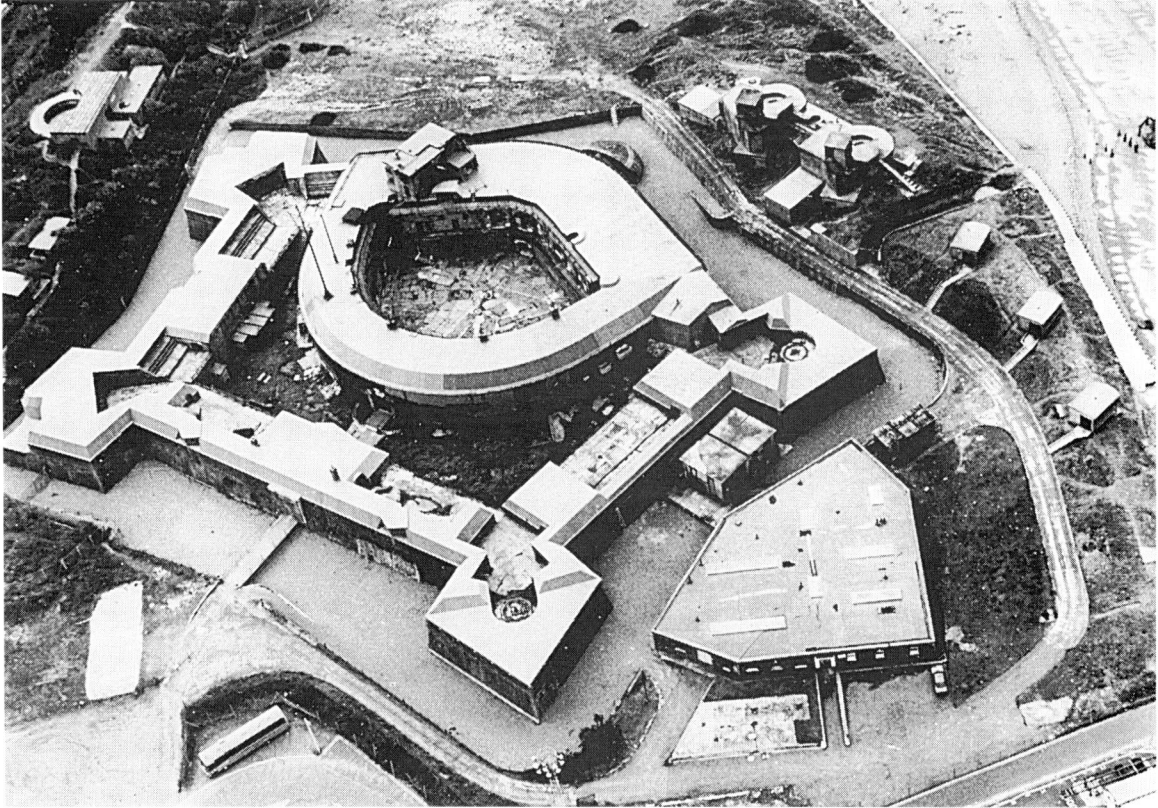

Figure 20 Aerial view of Landguard Fort, Suffolk, an example of earlier fortifications being reused and their role redefined in later conflicts. This site is now open to the public. Recent excavations were conducted in advance of it becoming an improved visitor attraction (Photograph: Roger J C Thomas)

Figure 21 A lozenge shaped pillbox at Spurn Point, East Riding (Photograph: Roger J C Thomas)

plugged by concrete anti-tank obstacles. At points along the stop line, at road, rail, and river crossings, pillboxes and anti-tank gun emplacements would be positioned. The aim was to halt the enemy armour, and then destroy it. Bridges and other structures that might be utilised by the enemy were prepared for demolition (see Dobinson 1996c for details). Associated with these measures, the development of petroleum warfare systems included sea flame barrages, burning beach installations together with flame fougasses, and fixed flame defences inland.

As part of a system of in-depth defence, individual villages and towns were prepared for all-round defence. Vulnerable points, and installations, such as anti-aircraft batteries, radar stations, decoys, coast batteries and searchlights, were generally defended by light pillboxes or sandbagged weapons positions. All open areas with a length of 500 yards or more (often including straight stretches of road) within five miles of the coast, or an airfield or other vulnerable point, were blocked with a variety of different types of anti-landing obstacles (machine- or hand-dug trenches, poles, cables, tree trunks, old cars, piles of rubble etc). An example of trenches survives within the Sutton Hoo Saxon burial ground in Suffolk.

Auxiliary unit counter defence was Britain's secret resistance army whose purpose was to harry the German lines and communications in the event of a landing, and act as the nucleus for future resistance. The main structures here were underground operational bases (OBs) or 'hides', usually in the form of a buried room with escape tunnels, and the provisions and furniture necessary for their occupation, as well as radio stations and stores. In addition to the Auxiliaries, the British Resistance Organisation included a separate network of trained agents with a message-passing system linked back to the military through hidden wireless stations and underground relay stations.

It has been estimated that, by October 1940, some 28,000 pillboxes and anti-tank gun emplacements had been built or were under construction in Britain, and 1500 miles of anti-tank obstacles (concrete obstacles and artificial and natural-improved anti-tank ditches) were planned. Removal of these categories of defence works was begun before the end of the war.

[OBJECTIVES: A13, B2, B4, C2–4, D1–3, E2, E4–5, F]

Coast artillery

The use of fixed artillery to protect the coast from hostile ships is one of the oldest practices in the history of England's defences (Saunders 1989). From the fifteenth until the second half of the twentieth century, coast artillery provided home security as well as protecting communications and trade networks across Britain's Empire. During this time batteries of fixed guns formed the first line of defence for the navy's anchorages and the larger commercial ports. Apart from a brief period early in World War II, when improvised batteries formed a continuous cordon around the coast, England's modern stock of coast artillery sites was dominated by positions originating before 1900. Coast artillery was finally stood down in 1956.

There were four classes of twentieth-century coastal batteries (Dobinson 1999b):

- anti-motor torpedo boat batteries
- defended ports, which include counter-bombardment, Close Defence, and Quick-firing batteries
- emergency batteries of World War II
- temporary and mobile artillery

Associated with these defensive positions were anti-shipping and submarine nets and booms.

Unlike some other classes of military monuments, coast batteries display considerable variation in type according to: construction date and the use of earlier fortifications; the types of gun housed on these sites; and their precise function. Nevertheless these four monument classes do have characterising features which make them identifiable in the field.

For coast batteries the same three stage methodology was adopted for assessing sites in England as was used for air defence (above). In all, primary sources confirm that 286 locations in England were occupied by 301 separate batteries in the period 1900–56, many of which made use of earlier fortifications of 1660–1900. These earlier fortifications were the subject of a separate earlier evaluation by MPP. Six Monument Class Descriptions were written for these earlier fortifications, and a desk-based assessment of sites in England was undertaken by Andrew Saunders. Many of the coast batteries identified as nationally important were already scheduled, and have been subsequently reviewed under MPP along with the earlier fortifications in which they were sited. For comparison, 235 sites are recorded for Scotland, 54 for Wales and five for Northern Ireland (Redfern 1998a, 3).

In England, aerial photographic assessment has shown that at least 115 of the 286 twentieth-century coast batteries (40%) have been removed, the great majority by human agency through agriculture and (on the east coast especially) coast erosion; this leaves some 60% of sites surviving in some form (for more detail see Schofield 2002a, 277–9). This comparatively high level of survival for coast batteries (compared to decoys and anti-aircraft sites for example) is due largely to their use of earlier 'historic' fortifications. For instance, 81% of sites constructed prior to 1921 have survived, compared with only 39% of 'new' sites constructed in the period 1938–45.

[OBJECTIVES: A13, B2, B4–5, B12, C2–4, D1–3, E2, E4–5, F]

Figure 22   The remains of a coast battery in East Yorkshire recently fragmented by coastal erosion (Photograph: John Schofield)

Aviation

Airfields

Powered flight, and in particular its application to military purposes, has had profound impact on the human experience of the twentieth century and on the modern landscape, and military airfields represent the most significant manifestation of that impact (Lake 2002). Military airfields are typically extensive and complex sites, whose planners took into account the functions of a technology-based service and the accommodation, ordered by rank, of communities of flyers, technicians, administrators and their families. They were built in great numbers: about 250 flying stations were in existence in summer 1918, most of which were subsequently abandoned; approaching 100 built in permanent fabric between 1923 and 1939; and the country's total of 150 expanded to 740 during World War II. There is a wide functional range of site types, from operational (bombers and fighters, maritime aircraft, army cooperation, air mobility and reconnaissance) to the purposes of training (see above) and the storage of reserve aircraft.

Britain's first military fliers were balloonists, and from the 1870s the Balloon Section of the Royal Engineers worked from Woolwich, Chatham, and Aldershot. Aldershot led to Farnborough, where the Balloon Factory opened in 1905 to build Britain's

first airship. Aeroplane training began in 1910 at Eastchurch (naval) and Larkhill (army), but it was only with the RFC's foundation in 1912 that it gained a formal stamp. Upavon, the Central Flying School, opened in that year, and before the Great War it had been joined by operational bases at Netheravon, Montrose and Gosport, and at coastal locations for seaplanes. Then came the war, and the growth of aeroplane and airship stations for defence, training, storage, reconnaissance, maintenance – the full range of air power functions and needs. It was this mass of sites that the RAF inherited in 1918 – and promptly cut.

From 1923, when the first phase of inter-war expansion commenced, air bases were built in permanent materials (mostly brick and concrete) and planned on dispersed principles, the result being the planning of hangars on arcs and mess buildings in linked compartments in order to minimise losses to machines and personnel. The improved design of post-1933 buildings was a product of the Government's request – spurred by popular fears over rearmament and the impact of air bases in the countryside – for the Air Ministry to liaise with the Royal Fine Art Commission over the matter of station design. Planting schemes became a significant part in the design of air bases. The gas decontamination centres and protected operations blocks which

RAF Hullavington: Site Plan 1946
Flying Training School & Aircraft Storage Unit
Based on Air Ministry Drawing 2765/46

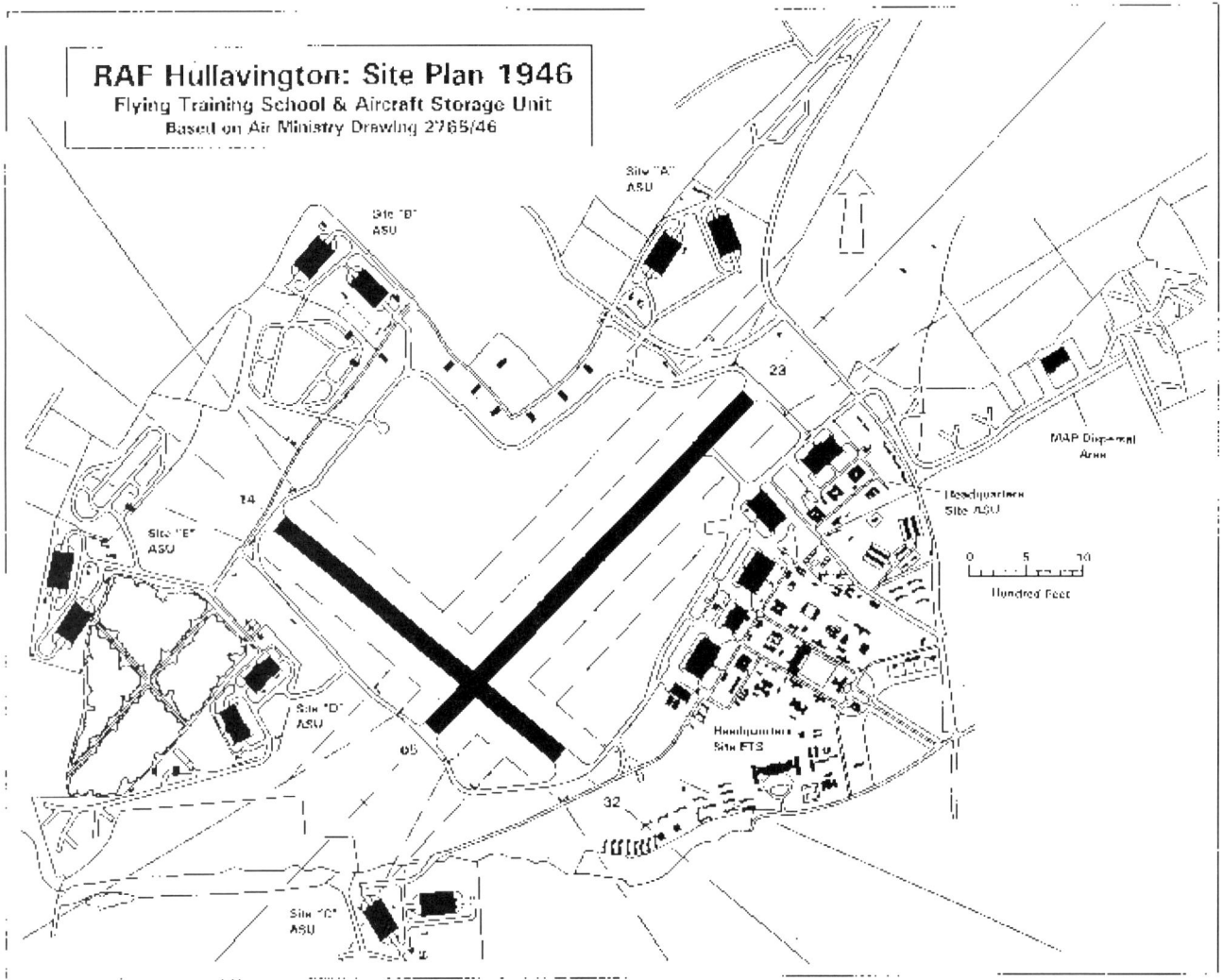

Figure 23   Plan of Hullavington, Wiltshire. The site embodies the improved architectural quality associated with the post-1934 expansion period of the Royal Air Force. Its flying field remains, bounded by planets of hangars built for the aircraft storage unit in 1938–39. (Drawing: Paul Francis)

appeared on Royal Air Force bases from 1937, along with the flat roofs widely introduced in the same period, were designed to counter the effects of incendiary bombs and bomb fragmentation (Francis 1996, 186–193).

Airfield size is closely related to technological development. A survey of Lincolnshire airfields established that whereas the average size of airfields in the 1914–18 period was 167 acres, this had increased during the 1930s to 400 acres and by 1945 to 640 acres (Blake 1984, 210). In the second half of the 1930s, increasing attention was being given to the dispersal and shelter of aircraft from attack, ensuring serviceable landing and take-off areas, and the control of movement through the availability of wireless. The result of the latter was the development of the control tower, while the planning of protected hangars in the aircraft storage units constructed from 1936 arose from developments in dispersal policy. The first airfields with runways and perimeter tracks were introduced in 1938 to ensure all-weather serviceability in an era when aircraft

were becoming heavier and the concurrent adoption of retractable undercarriages with small wheels and high pressure tyres made grass increasingly difficult. These airfields were concentrated in Fighter Command, particularly in 11 Group in south-east England. A major feature of the period 1942–45 was the widespread construction of airfields with concrete runways and hardstandings for the perimeter dispersal of four-engined bombers (Betts 1996). Hangars were only built for servicing, the other buildings being sited amongst clutches of domestic and technical sites scattered across several square miles of surrounding countryside. The deployment of air bases in World War II also reflected key strategic considerations, from their siting in reaction to German occupation of north-west France, to the construction of bases in eastern England in support of the Strategic Bomber Offensive and the advanced landing grounds sited in southern England in support of the Allied invasion of north-west Europe.

Analysis and an assessment of what has survived in England, and the comparison of survival with original populations, have been undertaken by Paul Francis, author of *Military Airfield Architecture* (1996). Colin Dobinson has undertaken additional archival research, exploring themes relating to airfield planning and architecture, particularly from 1923 (Dobinson 1997). Based on these related programmes, key surviving sites, and recommendations for protection, were identified in a draft consultative report issued in May 2000 (Lake 2000). Criteria to support selection included: completeness; historical associations (eg with the Battle of Britain, see Lake and Schofield 1999); buildings with architectural or historic merit; and international context (for example Britain's role within NATO). As an example of survival, of the 450 control towers that existed in England in World War II, 220 survive in some form today.

[OBJECTIVES: A13, B2, B13, C2–4, D, E2, E4–5, F]

Airfield defences

Defence was integral to the planning of new airfields from the early 1920s, following which there were two main phases of design and construction (Dobinson 1998a). The first phase was part of the programme of airfield building designed to parallel Germany's increasing rearmament in the 1930s. At this stage defences were designed to provide protection from air attacks aimed at the destruction of fabric and equipment. Dispersed layouts, air-raid shelters, protected buildings and anti-aircraft guns were the principal measures used. The second phase followed the realisation in spring 1940 that airfields might be targets in a strategy aimed at capture. This phase is represented by the construction of pillboxes and Battle HQ buildings, from which defence of the airfield would have been coordinated. In the Cold War, NATO-funded Airfield Survival Measures (ASM) were implemented on certain British airfields. These included hardened aircraft shelters, hardened squadron operations facilities, runway control pits, communications and protected fuel installations. In addition, battle damage repair materials were stored nearby and engineer units trained in rapid runway repair. Airfield defence does not therefore represent a coherent group of related sites and structures, but a loose collection representing distinct and separate phenomena.

In England, World War II airfield defences have been assessed on the basis of three main criteria: those that survive on sites which are of key historic importance, and where the quality of survival of a range of airfield buildings and the flying field is exceptional (defences at six airfields are considered to be of national importance in this regard); those where the defence provision is largely complete (twelve airfields); and those where individual structures are rare survivals (such structures are currently recorded on 36 airfields).

The following figures give an indication of survival (after Francis, nd): for fighter pens, contemporary site plans show 696 examples in England (and not all airfields are represented in the plan series); only

some 40–50 examples are left on eleven airfields, representing the three main types. All twelve of the original fighter pens survive at Perranporth in Cornwall. Anti-aircraft gun towers are rare, four examples being recorded on airfields. Similarly, sleep shelters are very rare nationally, with six examples recorded. Finally, some 242 Picket Hamilton forts – a pneumatic pillbox designed specifically for airfield defence – are recorded as having been installed on 82 airfields (Dobinson 1998a); only nineteen are now recorded as surviving on twelve sites.

[OBJECTIVES: A3, A13, B4, B12, C2–4, D1, D3, E2, E4–5, F]

Crash sites

Although there is a history of flight extending back to early experiments in 1911–12, twentieth-century military activity over the UK reached its height during World War II, with aircraft of the RAF, Fleet Air Arm, Luftwaffe, Regia Aeronautica, US Army Air Force and US Navy all operating over, around or from the British Isles. Following the Battle of Britain and the Blitz, from 1941 onwards the UK was the base from which strategic bombing, anti-U-boat operations, and offensive fighter sweeps were mounted, and from mid-1944 airborne and supply operations in support of the invasion of north-west Europe. The UK was also used for training on a significant scale. In the latter half of the twentieth century, encompassing the Cold War, military aviation in the UK was primarily centred upon defence and monitoring, to counter the Warsaw Pact, and provide the UK's independent nuclear deterrent in the form of the V-Bomber Force.

It is not surprising therefore that many aircraft have crashed in and around the UK, either in conflict or during training exercises. Research into those crash sites in England has been undertaken by MPP (English Heritage 2002; Holyoak 2002). Research has included surveys of:

Aircraft types, using secondary documentary sources with the intention of identifying numbers of aircraft originally produced, the number of military aircraft that crashed within the UK up to the end of World War II, numbers preserved within museums (these combining to determine the rarity of crashed examples) and historical context

Contemporary documentary sources relating to crash sites.

The MPP survey was restricted to the period 1912–45. Aircraft in the post-World War II era were generally produced in smaller quantities than those of either World War I or World War II, and examples of all major types survive. Their crash sites are thus considered to be of much less archaeological merit.

Initial consultations with the Royal Air Force Personnel Management Agency and the Air Historical Branch of the Ministry of Defence, the British Aviation Archaeological Council and the National Trust were conducted on these principles with a view to agreeing best practice in terms of managing the remains of crash sites.

The resulting assessment was intended to evaluate the nature and extent of the resource, its current management, and options for future conservation strategies. It estimated that of the 226 military aircraft types of all nationalities in use over the UK between 1912 and 1945, examples of only 92 (40%) of them are preserved in museums (Holyoak 2001). By period the figures break down into the following:

1912–1918 85 types of three nationalities (British, French, German) of which examples of only 21 (24.7%) are preserved in museums

1919–1936 48 types of two nationalities (British and French) of which examples of only 12 (24%) are preserved

1937–1945 93 types of four nationalities (British, German, US and Italian) of which examples of 59 (63.4%) are preserved.

Research for the MPP survey also identified a variety of data sources providing information on military aircraft crash sites. These are described elsewhere as are the relative archaeological merits of crash sites (English Heritage 2002).

[OBJECTIVES: A3, B2–3, B12, C1–4, D1, D3, E2, E4–5, F]

Cold War

Chronologically the Cold War spanned a longer period than that from the outbreak of World War I in 1914 to the end of World War II in 1945. During much of this period membership of NATO is of fundamental importance, and certainly for the later years of the Cold War it is important to differentiate between: NATO-funded infrastructure; UK infrastructure supporting forces assigned to the NATO Central Region in Germany; and infrastructure provided by the UK for foreign NATO forces. These distinctions are fundamental to understanding the fabric that survives, and provide a useful framework for characterising most Cold War material culture. But equally, construction work can be seen in terms of two clear phases, corresponding to the two periods of heightened tension between the superpowers, known sometimes as the First and Second Cold Wars, and it is this framework that forms the basis for English Heritage's study of Cold War sites in England. The first period runs from the outbreak of the Korean War in 1950, through the massive rearmament programme which followed, until the early 1960s. The Second Cold War was the period from the late 1970s, with worsening relations between East and West and a massive armaments spending programme led by the United States. The end of the Cold War is taken to correspond to the fall of the Berlin Wall in November 1989.

The archaeological recording and assessment of Cold War monuments in England has followed a slightly different pattern to that of other monument classes.

Figure 24 Little now survives of this crashed Heinkel 111 on Lundy, Bristol Channel (Photograph: John Schofield)

Figure 25   Former RAF Alconbury, Cambridgeshire. 'The Magic Mountain', this massive, double-storey bunker was built during the late 1980s primarily to process and analyse data from high-altitude reconnaissance aircraft. It represents the pinnacle of Cold War bunker architecture (© English Heritage, AA023746)

Documentary research at the National Archives concentrated on just four topics: 1950s Rotor radar sites, Bloodhound surface to air missiles, Thor missiles, and Royal Observer Corps warning and monitoring posts (Dobinson 1998b). In 1997 the RCHME initiated a field-based project to record the monuments of the Cold War, also including limited documentary research and aerial photographic recording (Cocroft and Thomas 2003). This project continued after merger with English Heritage in April 1999. Information from this project, the work by Colin Dobinson, and analysis of aerial photographs by Mike Anderton (2000) was subsequently used by the MPP (Cocroft 2001; Cocroft 2003; James 2002) to identify which Cold War structures had particular significance, and which should be recommended for protection. Up-to-date information was also sought from the Internet, in particular the pages hosted by Subterranea Britannica (www.subbrit.org.uk/rsg/index.shtml), though it is recognised that much remains unknown to researchers outside military, NATO and Home Office circles.

In England, the assessment process divided the monuments into nine Categories, which were then subdivided into thirty-one Groups, and further if necessary into Classes and Types. In many Groups the original populations were small, often comprising not more than ten or twenty sites. Despite their relatively recent date, rates of loss have been high. For example, of the eleven Bloodhound Mk I surface to air missile sites operational between 1958 and 1964, only two survive in near complete condition. Thor intermediate range missiles were deployed on twenty sites between 1958 and 1963. Of these six have been totally cleared and only four sites retain a near complete set of component features. Of the most numerous monument Class, Royal Observer Corps Underground Monitoring Posts, 1026 of which were built in England, fieldwork (mainly by members of Subterranea Britannica) has confirmed that about 30% have been demolished. The Cold War was also characterised by many sites which carried out unique functions, such as the Ballistic Missile Early Warning

Figure 26   Hardware, military vehicles, ships and aircraft draw many visitors to military museums and have contributed much to the enthusiasm that now exists for the buildings and monuments from which they operated (Photograph: John Schofield)

Figure 27   A mock SA-6 'Gainful' missile launcher, at RAF Spadeadam, Cumbria (Photograph: Roger J C Thomas)

Station (BMEWS) at Fylingdales, North Yorkshire, or the less well-known NATO Forward Scatter communications stations of the NATO Allied Command Europe 'ACE HIGH' network (eg at Stenigot, Lincolnshire), both of which are now demolished. Operational sites such as these, but also structures found within Research and Development establishments, require individual appraisals of their strategic and technological significance. The assessment process also highlighted areas where knowledge is still lacking.

Although included in the English Cold War assessment, the material remains of the peace movement are being studied separately (eg Schofield and Anderton 2001) to provide a balanced account of the Cold War period. This work gives the Cold War relevance in considering social inclusion and cultural diversity agendas. Wall art on Cold War bases is also a significant consideration (Cocroft and Schofield 2003).

[OBJECTIVES: A2–4, A6–7, A9–11, A13, B1, B3, B6–7, B9–14, C, D, E, F]

Plant, hardware and military vehicles

Nearly all twentieth-century military sites, as encountered today, are skeletons of their former selves, and although stripped of much of their equipment they will often contain the only physical evidence of complex weapons systems. In this category we might include such recent sites as the 1980s NATO Ground Launched Cruise Missile (GLCM) sites with cruise missile shelters at Greenham Common and Molesworth, where most of the missiles and their launchers were destroyed under treaty obligations.

Before closure of most military sites, equipment and hardware is usually removed and often scrapped. Sites that retain original equipment are rare and important, illustrating the relationship between technology, the architecture and operation of the site or structure. In such cases it is important that the associated documentation is also located and preserved.

The interpretative and educational value of a site will be greatly enhanced if examples of contemporary plant, equipment or vehicles are available for display alongside or within the structures from which they operated. This association will be of particular importance if a historic association between the object and the site can be established. However, where examples of original equipment do survive, in most cases due to there being no significant security concerns, it is generally impossible to preserve this association in the longer term, given conservation or health and safety considerations. Where equipment, boats and ships, aircraft and vehicles have been removed to museums, few attempts are then made to explain how they operated in relation to the site or structure from which they came.

One of the main innovations in the conduct of warfare in the twentieth century was the development of the internal combustion engine and the resulting mechanisation of conflict (coincident with which was a reduction in use of the horse, although horses remained in both military and civilian use well beyond the end of World War II, and much infrastructure remains). Numerous examples of twentieth-century military hardware that reflect this development, including aircraft and vehicles, are preserved by museums and in private collections, including within barracks and military teaching collections. However, not all are of equal historic importance. Many have been rescued from dereliction and are heavily restored; some are regularly operated requiring frequent replacement of components. In so doing they fulfil a valuable educational role in stimulating interest. There is also a small number of aircraft, boats, ships, vehicles and other objects which have a close association with an historical event or person. These objects have an individual significance in the narrative of war. In some instances a poorly preserved excavated aircraft or vehicle will have more significance than an object taken straight from a store to pristine museum display. Other military artefacts are examples of major technological and scientific advance. Even where they failed to enter service, often for political reasons or obsolescence through changes in threat, they have nevertheless often brought significant industrial benefits.

What is lacking is a corpus of surviving military plant, hardware and vehicles that are of historic interest, and a correlation between that and a list of extant and publicly accessible sites to which they belonged. The military museums have an important role in this activity. With such a correlation it should prove possible to assess display and interpretative options. Enough specialist groups exist to make this corpus realistic. The work on military aircraft crash sites is an example of what can be achieved (English Heritage 2002, and above).

Obsolete vehicles due for destruction under treaty obligations and the equipment of former adversaries no longer needed for evaluation purposes have often ended their days as range targets. A few of these are now rare survivals of their type.

[OBJECTIVES: A3, A7–9, A11, B2–3, B6, B12, B15, C2–4, E5, F]

Theme 5: Commemoration

War memorials and commemorative sites

War memorials stand at the heart of almost every community in Britain, and range in scale from wooden commemorative panels in parish churches to spectacular public architectural and sculptural ensembles. It is estimated that there are about 50,000 war memorials of all sorts nationally, of which over 8,000 are freestanding structures. War memorials also vary considerably in terms of their structural stability and levels of care and maintenance. While many are in excellent condition, others are taken for granted and neglected, and in some cases are subject to loss and

Figure 28   War memorials: a common feature in Britain, are important for memory and sense of place (Photograph: John Schofield)

Figure 29   The control tower at East Kirkby, Lincolnshire. This control tower has been restored as the centrepiece of a museum that commemorates the history of this airfield, opened in 1943 in support of Bomber Command's offensive. It was one of 160 Watch Office for All Commands control tower buildings constructed, of which 82 survive (Photograph: Jeremy Lake)

damage through vandalism. The Friends of War Memorials is a charity set up in 1996 to draw attention to the importance of war memorials and to the various threats they face.

Most war memorials commemorate the casualties from the two world wars. The dead of later wars, such as the Korean War, usually only appear – literally – as a footnote on the bases of earlier monuments. There are also now memorials to specific groups of war dead including – controversially – those shot for cowardice during World War I. In Northern Ireland there are memorials to those who died in the Troubles. Some memorials take the form of stained glass windows in churches.

As well as formal war memorials, many of the wartime sites described above and many places from which such structures have already been removed will provide for someone, or for a particular group with a common identity, a place of memory; a place of enspiritment (Read 1996). This may be for former occupants, employees or for the local community. It may be for those who played 'war games' in these sites as children. All places which represent some aspect of war and conflict in the twentieth century are likely to have a value in these terms, for belonging, commemoration or remembrance and sense of place (Schofield 2002b). Some trophies and objects, including works of art survive on airfields, such as the guns, aircraft parts and lace memorial cloth at RAF Coltishall. As airfields are closed, such items are vulnerable to dispersal.

[OBJECTIVES: C2, C4, D4, E4–5, F]

War cemeteries

During the twentieth century, many British, allied and enemy service personnel lost their lives within the UK and were buried locally. In some cases their remains have been removed to war cemeteries but others remain in local churchyards and cemeteries. There is an increasing interest from across the Atlantic in locating the individual war graves in cemeteries throughout the UK of Canadian and US service personnel. Over 60,000 British civilians lost their lives in World War II through enemy action. In some cases they were buried in mass communal graves; for example, the 400 civilians who died during the Baedeker Raids on Bath on 25–26 April 1942. Other single graves may mark, for example, the first Home Guardsman to lose his life on duty in the area. The importance of such cemeteries and individual graves brings home to future generations the sacrifices that, not only service personnel, but ordinary civilians of all ages made during the wars

Figure 30 The interior of a hangar at the Imperial War Museum, Duxford. Duxford has retained the best preserved fabric of a World War I airfield, including three paired hangars (Photograph: Mike Williams © English Heritage)

and the impact that such deaths must have brought on families and communities.

[OBJECTIVES: C2, C4, D4, E4–5, F]

Military museums

The role of museums in preserving artefacts has already been mentioned. For the army there are approximately 150 Regimental and Corps Museums across the UK. These vary enormously in size and style from those housed in prestigious castles to small collections in Territorial Army Centres. Some of these buildings will warrant preservation on their historic or architectural value alone, while others are state-of-the-art museums, which in due course may deserve similar treatment. In addition, there are sites ranging from the Imperial War Museum Cabinet Office War Room in London to the abandoned Cold War Regional Government Headquarters (RGHQ) buildings now open to the public. The growth of such military museums shows on one side the wish of the armed forces and indeed ex-service personnel to record and preserve their heritage; and on the other side the growing interest of the general public in military heritage. These military museums are particularly important as they record the lives of individual service people in the units associated with them. They also form a focus for those interested in the memories of the survivors of past conflicts. In addition, many of these museums have extensive records and libraries which are invaluable to researchers.

[OBJECTIVES: C2, C4, D4, E4–5, F]

Part 2: Research agenda

What follows is an agenda, divided according to six themes and then – within those themes – topics. Sometimes generic, sometimes specific, these are areas where research is needed to further understanding and ensure that informed conservation is the underlying principle for managing change, and to promote public awareness and enjoyment of military heritage. It is important to stress the various and diverse means by which these objectives can be met, though there are key areas for cooperation and engagement, notably with the public and local enthusiast groups, and with the university sector. This agenda is indicative, however, and stands only as a first attempt to prioritise and focus research into recent military heritage. As an initial agenda this document should not preclude the consideration or resourcing of other currently unformed research objectives and topics that may have equal or even greater significance and worth to those identified here. This agenda represents a starting position; nothing more. The six themes are as follows:

A – Improve understanding of the built resource

Objective: to continue to investigate what was built, where and when, and what form the sites took, using appropriate sources (eg documents, field remains, aerial photography).

Much is known of what was built, where, when and why. Indeed this is the area in which most significant progress has been made over the last decade. This knowledge contributes to public awareness and understanding of how the landscape was transformed and fortified; how Britain prepared and was mobilised for war in the period 1914–89. An Atlas of Britain at War would be one result of this research, as would publication of some of this information over the Internet. But some significant gaps in our knowledge do remain.

Specific areas for research:

A1 – D-Day preparations and support

While much is known about the embarkation for D-Day, and the construction and maintenance tasks pertaining to the fleet and artificial harbours, less is known about the build-up to embarkation: the camps; the communications; training; logistic installations and so on. As with other aspects of the operation, archives exist which cover these areas. These have now been located yet remain to be studied systematically.

As one of the most significant historical events in recent history, understanding the build-up to and preparations for D-Day is an important task. The archives exist in the form of maps at the MoD Map Library and at the National Archives (Public Record Office), while wartime aerial photographs will also be useful source material. These should form the basis of systematic study. As only a few sites involved in these preparations will survive, and these only as ephemeral and largely buried remains, an examination of where the sites were – and what survives (Theme B, below) – is a priority, as is an impression of their impact in the countryside and on landscape change.

A2 – Camps (general)

Research is required to better understand the form, distribution and function of camps. There is a need both to collate documentary sources for camp locations and types, as well as searching secondary sources for information on construction, layout and function. An overarching survey will also need to take account of the built heritage as well as earlier and transient camps surviving as buried remains and earthworks. The impacts of camps and their inhabitants on the surrounding landscape and on urban areas will be of social historic interest and should also be addressed in any national study.

A self-contained and coherent research project could use National Archives sources to determine site locations and secondary sources for contextual information, followed by analysis of large-scale post-war Ordnance Survey mapping and aerial photographs to locate sites and assess modern survival (see below). Specific work on D-Day [A1] and PoW camps [B8] is needed, and this is documented elsewhere. Work on TA centres – now underway – based on field visit, and documentary sources, should be accelerated to completion [see B9 for details].

A3 – Cold War

Despite the work undertaken to date (Cocroft 2001; Cocroft and Thomas 2003) gaps remain in our understanding of Cold War material culture. These gaps can be filled in different ways, though some basic principles apply in most cases, such as the value of gathering information from archive and testimonial sources. Communications-related sites, R&D and production sites relating to key programmes of research, civil defence, training , electronic warfare and the role of Information Technology are some key areas where

Figure 31   Former RAF Greenham Common, West Berkshire. The draw-bridge like door of one of the shelters used to house the cruise missile launchers during the 1980s (© English Heritage, AA000532)

basic information on what was built, where and why is needed.

This objective can be achieved through targeted research, building on official histories, oral testimonial evidence and archives where available, to provide a characterisation and typological framework and indicative site lists.

A4 – Civil infrastructure

For both World War II and the Cold War there was a vast infrastructure embracing buffer depots, food stores, strategic material stores, fuel stores and transport, and questions remain about how this infrastructure was built and used, and how essential resources were mobilised. This is a large subject which remains to be explored, and will be essential to understanding Britain's preparations for war, how Britain continued to maintain basic functions during World War II, and how civilian and military logistic operations would have been maintained in post-attack environments during the Cold War.

Archive surveys, and the study of secondary sources and personal testimonies, could combine to produce a detailed insight into infrastructural operations both during World War II and the Cold War. This

probably represents two separate though related studies, one for World War II and another for the Cold War period. A basic task will be to compile site listings by type for each period. Completing this typology and an initial gazetteer must be the priority, and is a necessary precursor to any further work.

A5 – Searchlight emplacements and barrage balloon sites

As one of the enduring images of the Blitz, searchlights form much of the basis for collective memory, along with sirens and shelters. Although a sample list of documented searchlight emplacements in England has been produced (Dobinson 1999c), a definitive distribution of searchlight emplacements and of barrage balloon sites is not yet available. Some attempt to determine where and how many sites there were would be useful for purposes of interpretation and public education at a local level. Searchlight policy, site form and their evolution over time were explored in the anti-aircraft volume of the English Heritage *Monuments of War* series (Dobinson 2001).

Archive sources exist in the National Archives documenting site locations. Many of the military area defence schemes include complete lists of searchlight

Figure 32   Army border road block in County Fermanagh, dismantled in 1999 as part of the ongoing peace process in Northern Ireland. (© Environment and Heritage Service, Northern Ireland)

sites with wartime Cassini grid references; there are also German aerial reconnaissance maps which show their distribution. These sources could be the subject of a concerted national study, using SMRs and Defence of Britain Project records to confirm some surviving examples. Some plough-levelled searchlight emplacements have in the past been interpreted as the remains of prehistoric burial monuments. A correspondence analysis between SMR entries and documented emplacements could also have the benefit of removing any ambiguity.

A6 – Internal security and aid to the civil power

Work on military heritage has so far concentrated on defence against external threats, principally from Germany in World Wars I and II, and the Soviet Union in the Cold War. Fortification in support of internal security has been largely overlooked. The army's peacetime duties in the support of aid to the UK civil power have been in Northern Ireland, but they also played an important role in securing vital national assets during the 1926 General Strike, and civil government contingency planning allowed for similar preparations in the event of comparable disputes in the longer term. An assessment of the scale and character of the 1926 operations would be a useful next stage in exploring this subject. Studies into the role of Military Aid to the Civil Community (MACC)

would be a helpful addition, especially in areas where remains will survive: flood relief for example, temporary prisons, anti-terrorist operations, supporting fire strikes (with Cold War era Green Goddesses), RAF Mountain Rescue, and coming up to date, support during the foot-and-mouth outbreak.

A particular initiative is needed in Northern Ireland, to continue to record military installations of the 'Troubles' period. Currently photographic survey is carried out prior to the demolitions that form an integral part of the peace process, and border checkpoints, hilltop observation posts and fortified barracks and police stations have been recorded to date. This programme of recording, and some attempt at synthesis, are necessary and should be considered priorities. In this respect, lessons may be learnt from the reunification of Germany when the hated 'Inner German Border' watch towers, barbed wire and minefields disappeared almost without trace within months of the Wall coming down.

Documentary sources bearing upon these matters exist, though some will inevitably remain closed to inspection. Archaeology therefore seems the most likely source of evidence to examine any surviving evidence of fortified factories, munitions works and so on. In addition to the specific needs of recording work in Northern Ireland, an initial characterisation study would be helpful for the UK, after which more detailed work can follow dependent upon research needs and where a particular and urgent threat has been identified.

A7 – Intelligence infrastructure

A survey is needed to identify, characterise and list the major specialist installations serving military intelligence in Britain throughout the twentieth century. Information on locations of the major sites is readily available in published sources, together with many of the smaller installations, whose whereabouts and form for the first half of the century at least can also be studied through primary documents. These include infrastructure for signals intelligence and intercept, together with buildings connected with code breaking, cryptography, air photo interpretation and intelligence activities including Y service and training. A separate study should include sites connected with the Special Operations Executive (SOE) and the British Resistance Organisation (BRO) from 1940–45.

This study – incorporating four distinct subjects (intelligence, electronic warfare, communications and special operations) – could be achieved through the published literature and – where available – from archives held at the National Archives (PRO). For the BRO, oral-historical evidence will provide supplementary and site specific information that is unlikely to survive in the very limited written records that were produced at the time. Combined with these sources an indication of surviving sites could also be swiftly obtained from follow-up fieldwork perhaps alongside map- and aerial photograph-based study. For the BRO, oral-historical evidence should assist with the location of poorly understood site types, notably: Operations Bases, stores, intelligence-gathering dead letter boxes, and the well-hidden radio stations and special relay sites.

A8 – Air navigation aids, 1939–45

As a follow-up to recent work on radar, a listing of ground stations for air navigation in World War II (Gee, Oboe etc), with characterisation and assessment, can be obtained from archives and follow-up fieldwork. These stations were significant in serving the Combined Bomber Offensive in the later war years in particular. Many of these stations were co-located with radar stations but others were not and an assessment of the complete group is required.

This study could be achieved through the published literature and from archives held at the National Archives. An indication of surviving sites could also be swiftly obtained from follow-up fieldwork combined with map- and aerial photograph-based study.

A9 – BBC wartime and Cold War broadcasting

There are many structures relating to BBC broadcasting, some of which (in the Cold War) were hardened: relay stations for example. It is likely that there are many sites of this type across the BBC and it is an area little investigated and poorly understood (but cf Martin 2002).

A rapid survey of archives, and published literature could produce an indicative list of structures associated with wartime and Cold War broadcasting.

A10 – Royal dockyards

Research priorities for the royal dockyards need to concentrate on identifying the precise locations of workshops and laboratories that may have played key roles, for example in the development of naval aviation, range-finders, radar, early ship-board missiles, the use of new materials such as glass and carbon fibres, hull-forms, steam, diesel and nuclear propulsion and quiet propellers. Despite recent characterisation work at Portsmouth and Devonport much also remains to be done to track the yards' growth and development, especially for the twentieth century. The architecture of naval aviation, including airships and hovercraft, is the subject of current research by English Heritage. The identification of buildings – and their naval purposes – requisitioned or constructed for wartime use, would give a greater appreciation of the economic, social and technological impact of twentieth-century total naval war.

A research programme is needed using archive sources, held locally and at the National Archives, to identify and assess key structures and their respective roles in evolving technology and research. The recently completed characterisation projects at Portsmouth and Devonport are a useful starting point, and consideration should now be given to completing that exercise for the twentieth century.

A11 Warships and submarines

Research is needed into the construction, development and design, fitting-out (including de-gaussing), deployment, maintenance, and decommissioning of warships and submarines. This should include a study of trawlers armed for combat.

An initial scoping study of current knowledge could be undertaken based on public records, published sources and oral testimonial evidence. Further research priorities can then be identified, and the value of surviving historic ships can be properly assessed, with the first stage perhaps a listing by class and type of vessels built, and identifying those that remain intact (whether in service or on display) and those surviving as wrecks in an archaeological context.

A12 – 'Z' batteries

Far more 'Z' batteries were constructed than have been identified in English Heritage's indicative review of documentary sources. The further study of location statements held at the National Archives would produce a full national distribution.

Further work at the National Archives is needed. Independently of that it is likely that a familiarity with the distinctive plan form of 'Z' batteries amongst those engaged in research and survey within Gun Defended Areas (eg staff engaged in English Heritage's National Mapping Programme) could lead to the discovery of further 'Z' batteries.

A13 – Overview

Bringing all of these various objectives together, and perhaps with those identified under Theme B (improve understanding of surviving resources), one valuable and popular outcome would be an Atlas of Britain at War, using the materials collected from archive-based research and field recording programmes to show – graphically – what was built and where. In addition to the national maps could be various local and thematic case studies to highlight the scale of militarisation and its impact on the local environment at various points throughout the twentieth century.

Once a significant number of these key projects is completed, a small steering group should be established to explore this option, and to examine contents, scope etc. The Atlas would have to be UK-wide (to the extent of the 12nm Territorial Limit) to be fully effective. An Atlas could also attempt to convey the degree to which some key strategic military sites were 'recycled' for different roles in separate conflicts. The time depth is something an Atlas could effectively convey, enhanced by an accompanying CD Rom or webpage.

B – Improve understanding of surviving resources

Objective: to continue the process of researching and documenting the surviving remains of sites and monuments of this period, whether through aerial, geophysical, remote sensing or field survey (including submerged environments), and at a national, regional or local scale.

While much work has been done, and significant progress made in this area, gaps remain, both geographic and thematic. There is a need to improve our understanding of survival and the reasons for it, in order to provide better public information, to integrate modern military sites more fully within conservation practice in the UK, and to improve our awareness of monument management and risk at a national level.

Specific areas for research:

B1 – Follow-up surveys for site types investigated under A, above

Follow-up surveys are needed for those classes identified above, where documentary and fieldwork projects will investigate original site distributions, to determine what survives and to what extent. These classes include: D-Day camps and marshalling areas etc; camps (general); infrastructure; searchlight emplacements and balloon barrages; internal security; intelligence infrastructure; air navigation aids, 1939–45; and BBC wartime and Cold War broadcasting and GPO sites, although for some of these the process of determining original numbers and location would at the same time reveal surviving examples. (Other specific aspects of Cold War are considered under B12, below.) The work on searchlight emplacements and balloon barrages is less of a priority: these classes contain many more examples of sites, all of similar form, and more therefore are likely to survive. Their general distribution can also be estimated from other related and well-documented monument classes.

Follow-up surveys would involve fieldwork and/or the use of contemporary aerial photographs to determine modern survival. For England aerial photographs held at the NMR would be an invaluable source, as they have been for analysing other monument classes in the past.

B2 – World War I

With a focus of attention on surviving remains of World War II and the Cold War, World War I has been somewhat neglected in recent assessment programmes (but see Schofield in press and chapters in the various Dobinson reports). Although it does feature as a component in several thematic studies (eg aviation, Lake 2000) and survey projects (eg Salisbury Plain Training Area, McOmish *et al* 2001), a synthesis is lacking, using archives to provide context and fieldwork to determine survival and assessment of factories, hospitals, PoW camps, defences, training areas etc.

A synthetic study of World War I on the home front, tapping into various English Heritage and other surveys undertaken to date would be a valuable exercise with a popular book a likely outcome. The objective would be to promote understanding through publication and dissemination, and to give a firm foundation to management decision-making. A separate study of World War I air defence would be a useful addition to current research.

B3 – Submerged archaeology

As well as terrestrial remains, much of the material culture of modern warfare survives underwater, off the British coasts or in lakes, rivers and other water bodies, both in the form of shipping (military and civil vessels representing convoys, raiders, minelayers, minesweepers, landing craft, submarines etc), aircraft, tanks, submerged cables and other D-Day artefacts, such as Mulberry harbour remains (see refs in Schofield 2001). The vast majority of these various forms of craft were cut up after each world war, and

Figure 33   A camouflaged pillbox at Harper's Gate, Leek, Staffordshire (Photograph: Roger J C Thomas)

other than a small number of craft still afloat, the few surviving monuments to the effort, organisation and bravery of some wartime populations are the vessels which sank, and the remains of those who drowned (Oxley 2002).

As an initial step, a study is needed that quantifies and characterises this resource within the wider context of submerged archaeology generally, and assesses its management needs.

B4 – Anti-invasion defences

Following the completion of the Defence of Britain Project, several areas for subsequent research have been recognised:

> A systematic study is needed of those documents at the National Archives containing defence schemes, which set out for each area the purpose of defence, and the location and type of its various components. A catalogue should be produced of those that survive, their geographical limits and the types of information each contains. This catalogue could usefully be made

available on the Internet, enabling research to proceed more effectively and efficiently. Here partnership with the National Archives could be fruitful, as would be a published catalogue and users' guide.

Having assessed the relevant defence scheme(s), the data acquired by the Defence of Britain Project, and the results of further fieldwork where appropriate or needed, can be undertaken with a view to further understanding defence policy and its implementation at a local level, and set within the national context identified and documented in Dobinson (1996c). The value of this type of study is being demonstrated by current detailed work on the defences of the Taunton Stop Line which includes reviewing military formations, their organisation, weapons and concepts of operations, together with doctrines, policies and procedures relevant to construction and garrisoning of the Stop Line (David Hunt pers comm). This is also the subject of study in a forthcoming volume in Dobinson's *Monuments of War* series.

More work needs to be done on the question of the removal of defence works both during and after World War II. This can be achieved through documentary sources for selected areas, and an analysis of aerial photographs from 1946–2000. A correlation of site survival or loss against post-war land-use will be a useful study in the context of the Monuments at Risk survey, and data collated for other classes of monument by English Heritage (eg Anderton and Schofield 1999).

More work is needed on the German aerial reconnaissance photographs, mapping, and associated documentation. This is a significant source, yet it is still not known exactly what was produced and what survives in archives in Britain, America, Germany, and possibly Russia (but cf Going 2002). As with the defence schemes this requires indexing with lists produced and made widely available of what survives where, and what the various collections contain.

Following the completion of English Heritage's defence areas work (Foot 2003), thought should be given to promoting further research and to where available resources are best placed. A self-contained research project to document and catalogue relevant sources, their location, content and scope, will be invaluable and will serve to promote much further research by the enthusiasts who have contributed so much already to this field of study. Further analysis of the Defence of Britain Project data could lend itself to research projects at dissertation or thesis level, or local studies. GIS-based work to combine strategic ideals and military doctrine with reality (through viewshed and field-of-fire analyses) is one possibility for dissertation work.

B5 – Anti-aircraft, bombing decoys, radar and coastal batteries

The work undertaken in England using modern aerial photographs to determine, at a national scale, what survives of the sites originally built, and how well they survive, could usefully be extended to cover the documented sites in Scotland, Wales and Northern Ireland.

This could be undertaken as a single study, with one researcher consulting sources in each of the three countries. Depending on the availability of suitable aerial photographs, this work could be completed within two years.

B6 – Utilities and communications

Measures were put in place during World War II to ensure the functioning of the railway system. Structures included hardened signal boxes, railway control centres and related air raid shelters. Little is known of this subject. Other utilities – electricity, gas, water etc – all had standby facilities, which continued into the Cold War. Subterranea Britan-

nica members have visited some of these sites, though little or no contextual research has been undertaken. Wireless communications, GPO and later BT, are a significant consideration in the UK and internationally, and require further research to outline development and use, and to record and assess surviving remains.

Research through archives and in particular local and national records may provide information about this. Information may also appear in the local defence schemes [B4, above]. Some official histories and archives held by GPO or BT, or the National Archives may assist with a wireless communications study. There is some urgency, particularly for recording railway buildings which are disappearing very rapidly due to rail improvements.

B7 – Training areas

Having established where the main training areas were, and what types of training they were used for (Dobinson 2000a), the next stage needs to be a characterisation project which aims to establish what remains at each of those training areas that characterises the activities that went on there, how significant those remains are in terms of the training area's primary function, and what should be retained.

This can be best achieved by looking at each training area individually, through archives, local knowledge and subsequent field visits. It is important that these findings are then included in any management plans and environmental statements drawn up for the training areas. The recently completed study of Okehampton Training area, and a current assessment of Otterburn exemplify the way forward. The Okehampton study was undertaken for Defence Estates by a private consultant; the Otterburn study was produced by an MPP Archaeologist (Tolan Smith, pers comm).

B8 – PoW Camps

With information on site typology and location already available (Hellen 1999), a subsequent project currently underway seeks to establish what survives and where of the purpose-built camps (Thomas 2003). Work is also needed to establish the significance and modern survival of those camps not built for the purpose. More specifically, work could usefully concentrate on those camps and other places that housed conscientious objectors and internees (e.g. Lloyd 2001), and recording the graffiti that is commonplace at surviving sites.

With purpose-built camps as the priority for research, aerial photographs and maps (for England, held at the NMR) can be used to determine what survives and in what condition. Once completed, work is needed on those camps not built for the purpose, and those that housed conscientious objectors. The recording of graffiti is also a priority, alongside oral-historical evidence where available,

Figure 34   Kelvedon Hatch, Essex. Many of the surface structures associated with the early 1950s Rotor radar programme were constructed in a local vernacular style, such as this generator building with its chapel-like appearance. Such designs offered a degree of camouflage, but may also reflect the continuation of pre-war concerns about the intrusion of RAF facilities in the rural landscape (© Crown Copyright. NMR, AA00/01060)

as is the impact of sites and their occupants on the local economy and society.

B9 – Territorial Army

The completion of the current national assessment of TA centres and drill halls is needed in order to assess and afford protection to and establish management principles for those surviving examples. The work is complete in some geographical areas, but this needs accelerating to completion. This study could usefully be extended to cover militia, yeomanry, volunteer and even TA and cadet ranges and training areas, as well as the Administrative TA Associated HQs (T&VRA), and TA Weekend Training Centres (WETC).

This project is already underway and producing good results. The format of the final reports should be addressed in order that they provide information in a way that will be helpful to national agencies and local planning authorities, and the work accelerated to completion perhaps with some limited resources made available. A popular book on the subject could encourage and enable local studies.

B10 – Civil Defence

To develop a better appreciation of the rarity and distribution of surviving civil defence structures, from individual air-raid shelters, to communal shelters, first-aid posts, wardens' posts, fire-watchers' posts, decontamination centres, control centres, rescue bases, 'Jim Crow' posts and training facilities etc. This should also cover the National Fire Service and post-war Auxiliary Fire Service including fire stations, headquarters, training facilities and emergency water

Figure 35 Mickey and Minnie Mouse on the walls at a prisoner of war camp in Lincolnshire (Photograph: Roger J C Thomas)

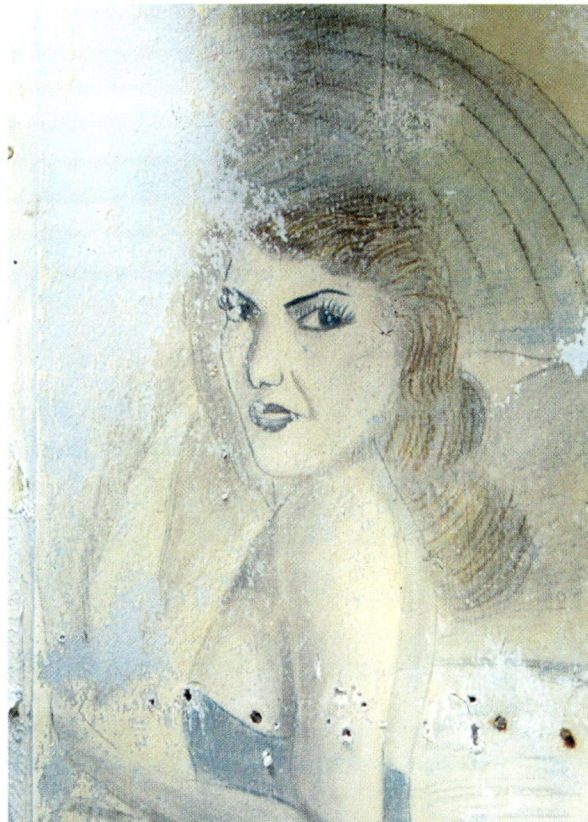

Figure 36 Woman's portrait at a prisoner of war camp in Lincolnshire (Photograph: Roger J C Thomas)

supply. Currently the typological range is understood, and there is some indication of survival stemming from local studies. But the degree to which these conclusions are widely representative needs to be understood, and an attempt to locate surviving examples is needed. The need to locate some privately built Cold War shelters is a priority in order that this class can be better appreciated, and its management needs recognised.

In the first instance this can be achieved through local studies, in a range of locations that can be seen as representative of the UK as a whole. Publicity from these local studies, and an emphasis on the significance of what remains, could then encourage further work elsewhere. This subject would be ideal for research by members of an enthusiast's group: Subterranea Britannica have already expressed interest, and Defence of Britain Project volunteers have recorded some examples. As has been demonstrated (eg Lacey 2002), this subject will also lend itself to undergraduate dissertation and other project work, where local sources can be compared to field remains, and oral testimonials.

B11 – Hospitals

Despite the completion of an English study of hospitals, which embraces military hospitals, gaps remain in our understanding. Specifically: post-war military hospitals, locations designated as Cold War Emergency Hospitals and USAF reserves.

Documentary research into the nature and form of these sites is needed, with subsequent site visits to assess the quality and condition of what survives, and determine their significance.

B12 – Cold War

Despite a survey and assessment of Cold War remains in England (Cocroft 2001; Cocroft and Thomas 2003) some gaps remain where little is known about monument types, surviving examples and their comparative completeness. Research is therefore needed in these areas, involving a combination of documentary research, aerial photographs and field checking. One subject includes those structures built to support the civil administration of the country in the event of nuclear attack. Structures in this category include local authority emergency and civil defence headquarters, protected operations centres built by the utilities and private nuclear shelters (see also B10). Detailed typological and locational information is also lacking for towers and masts associated with the government's emergency civil and military microwave communications networks, the NATO-funded systems, national systems, strategic communications, international radio and cable systems to communicate internationally with allies and the Home Office wireless and line radio systems.

A fundamental pillar of UK defence policy through most of the Cold War was membership of and contribution to NATO, which included land, sea and air forces, and this whole area – including command and control facilities, air defence systems, logistic installations and training facilities to support British forces assigned to the Central Region of NATO in Germany – requires further assessment. This can be addressed in terms of: infrastructure to support British forces assigned to NATO in the Central Region in Germany (Army 1st British Corps) and RAF Germany; Home Defence military infrastructure; and infrastructure to support world wide operations outside NATO.

The material culture of the 'peace movement' also needs to be better understood if the Cold War is to be presented in a balanced and objective way. Some work has already been undertaken on peace camps here and in the USA (See Schofield and Anderton 2000; www.lvrj.com [March 20 edition]), but more focussed research is needed to quantify this resource better, and understand its management needs.

National and local archive research, followed by fieldwork, is needed to identify the full range of structures used for post-war Civil Defence purposes and to ensure the due functioning of the utilities along with the location of private nuclear shelters. Knowledge of the latter topics may be found amongst the members of Subterranea Britannica. (See also B10). Further work, based initially on archive sources and oral-historical evidence, is needed also on communications and the UK's role in NATO. Work on the peace movement should be wide ranging, socially inclusive and innovative, exploring both the material remains and their future management needs.

Archive material is of considerable importance for the interpretation and management of sites at which Research and Development and production were undertaken, yet the release of documents to the National Archives will only partly answer questions on many sites. Many of the technical manuals and individual building drawings are regarded as too specialised for the National Archives or may have been destroyed for security reasons. To achieve a better understanding of the range of activities and functionality of key sites, it may be necessary to use oral history programmes to supplement the traditional historical record. The National Trust has, for example, started such a programme at the former Atomic Weapons Research Establishment at Orford Ness, Suffolk, and the British Rocketry Oral History Project forms a core of people pursuing this line of investigation. Some documentation, particularly relating to the United States Air Force, is held in the United States by the Department of the Air Force. However, this material will typically document unit histories, with only incidental references to infrastructure; some remains classified (see D3).

Achieving a better understanding of these R&D and production sites from the Cold War period could be achieved by targeted research, combining analytical field survey with searches for documentary and anecdotal evidence of specific sites or research/

production programmes. Where private firms were used in the production or research programme their own archives may provide valuable source material, while relevant documents are often found on site, even where the site has changed hands. It is likely that this area of research will be best followed through site specific research, rather than as a thematic programme.

B13 – Wall art

With brownfield sites being increasingly favoured for development there is a growing need to record wartime wall art and graffiti ('combat art' in American terminology), which falls into two distinct phases – World War II and Cold War, and on airbases also divides between British and American art. At one level a type of folk art, it also reflects the culture within different units at points in time, and the different cultures within different militaries. On PoW camps, paintings by German prisoners can be distinguished from those by Italians, for example.

Initially the production of a corpus of photographs of Cold War wall art currently held by the NMR for England is needed. Following that, additional unrecorded examples should be photographed and an interpretative exercise conducted, leading to publication. Given the development pressure at those Cold War sites where wall art is most likely to survive, this cataloguing and recording project, and subsequent analysis, is a priority. For World War II a catalogue of known examples of wall art is needed, from PoW camps, air raid shelters, mess buildings and so on. Thought also needs to be given to the conservation of wall art. A Guidance Note is needed, advising on best practice in recording and conservation.

B14 – The countryside at war

Although the countryside at war has attracted a good deal of historical writing, there is scope for more work on landscape change, for example through increased ploughing, deforestation, extraction of building materials. The impact of the Land Army is another consideration as is social displacement caused by the two world wars and after the end of the Cold War. This is represented by population shifts caused by military and labour service, continental and political upheavals which led to the permanent settlement of allied soldiers, prisoners of war and refugees. These pressures, along with the loss of housing due to bombing, led to many former military camps and sites being reoccupied, officially and unofficially, for civilian accommodation. This civilian reuse of military sites forms part of a wider topic of the taphonomy of the military estate, and the agencies that have combined to produce the assemblage of sites which survives today.

This subject can be approached both at a landscape scale, but also through archive-based and social historic research. This would be a suitable subject for historical geographers, and for related local history studies.

C – Pressures and perceptions

Objective: to determine and assess the various pressures that affect twentieth-century military remains, and changing perceptions of them.

A commonly held view is that public and professional interest in these recent sites is increasing concomitant with the increased pressure on sites for development and reuse. The research potential of these classes of monument – where we can establish (uniquely) precisely what was built and what survives – is also beginning to be realised. Research into these related areas would be of value, identifying future threats and benefits in this modern heritage, and examining its relevance for studying both the modern era, and lessons for understanding the more distant past.

Specific areas for research:

C1 – Crash site excavation and loss of records:

Although this principle may apply more widely, the focus here is crashed military aircraft which are typically subject to uncontrolled excavation and the subsequent loss of records. A Guidance Note, describing best practice, and for circulation to local planning authorities and aviation archaeology groups, has been produced to promote best practice in this area.

Following publication of the Guidance Note, a means to determine its effect and influence will be needed. This can include monitoring the quality of licence applications received by the Ministry of Defence, the number of licences granted, and recording the number of excavations notified to NMR and the local SMRs. Future actions will depend in part on how this Guidance Note is received by the aviation archaeology community for sites both on land and at sea. The British Aviation Archaeological Council can advise on this, perhaps two years after publication.

C2 – Stability and conservation:

As interest and awareness of these military sites increases, and more sites are preserved *in situ*, approaches to conservation and stabilisation need to be determined. Guidance Notes issued by the Ministry of Defence and English Heritage, the last specifically for airfield buildings (English Heritage 2003), will be relevant. However, for wider dissemi-

*Figure 37   Teachers are taught about World War II defences in an initiative to encourage
the use of recent military heritage sites in pursuing the national curriculum in England
(Photograph: Roger J C Thomas)*

nation and to cover the subject matter more generally, a further Guidance Note offering advice on a UK-wide basis may be appropriate, along with the provision of training, as part of building conservation training and appropriate educational courses. Some definitive studies of the stability of, and conservation issues presented by, such structures (including submerged military remains of all material types) will also be essential in determining future conservation strategies. Notable is the requirement to assess the conservation needs of (reinforced) concrete structures, and modern shipwrecks, and the conservation of wall art and outdoor camouflage, both *in situ* and establishing best practice for removal and conservation elsewhere (see section B13).

To meet these needs two research topics are needed: first, a dedicated piece of research (as part of a higher degree in building conservation perhaps) into the stability of concrete structures not built to last, and of the main conservation problems they present. A separate study is needed for wall art, and for shipwrecks. Second, the publication and dissemination of guidance to conservation staff on these related issues, and on the model of that published for Military Aircraft Crash Sites (English Heritage 2002).

C3 – Changing attitudes:

Attitudes to recent military sites (situated on land and underwater) have changed radically over the last two

decades. There is a greater popular interest now in these sites, both as historic places and components of the local scene. Heritage professionals also now regard these sites alongside more conventional monuments and remains, seeing them as opportunities and challenges to be addressed rather than as ugly, unstable and dangerous structures to be removed. Far fewer sites are now being removed without record than was the case a few years ago. In general, county archaeological staff and their equivalents in unitary authorities have a positive and constructive attitude to these remains, but that level of awareness has not yet been fully realised at district level. The subject is also not yet included routinely in undergraduate teaching despite its potential for addressing methodological, theoretical and interpretative goals. It is also not routinely addressed within the National Curriculum, although this too is changing (eg an initiative in Kent to teach teachers how best to use recent military sites, Barnes 2002). Research is needed into the role these twentieth-century military remains perform (why are they important and to whom?), and how they are best incorporated in community archaeology initiatives.

Research is needed into changing attitudes, following popular publications by English Heritage and others (English Heritage 1998; 2000). Have visitor numbers to military heritage attractions increased; and what attitude do local conservation staff take to proposals affecting modern military sites? What is the public perception? Is it generational? This general review of perceptions and

Figure 38   Anniversary service taking place at the Torquay D-Day slipways, June 2000. Present are members of the Normandy Veterans, a re-enactment group and local schoolchildren (Photograph: John Salvatore)

approaches could usefully form the subject of a dissertation or thesis by a student of heritage management and tourism, building on the results of the MORI poll undertaken within the context of Power of Place *(English Heritage 2000b).*

C4 – Social commemoration of warfare

All military sites have the capacity to evoke strong feelings and personal or community responses, and the way they may change over time are themselves important subjects for research. The cultural values attached to sites will inform decisions made about their future management and presentation.

Differing views of significance and the propriety of preservation go to the heart of the debate about what 'heritage' means. Heritage that is closer to us in time may stimulate fiercer passions and add a sharper edge to the issue that actually lies at the heart of conservation: the need to manage our environment so that it enhances the quality of our life while maintaining physical links for future generations to make fresh sense of their own past.

As above, this could usefully form the subject of dissertations or theses by students of heritage management and tourism.

D – Methodologies

Objective: to explore new approaches to modern military heritage and to ensure its integration with other related agenda, such as sustainability, social inclusion and the related fields of philosophy, sociology, geography and archaeological theory.

What, beyond the simple description of these sites, structures and landscapes, can be gleaned from

careful and detailed research, embracing perhaps archaeological survey and excavation, related oral history and archives and using other techniques borrowed from the humanities and social sciences? This theme is about developing the techniques to gain more from the subject and make it as effectively multidisciplinary as the subject matter requires. Many of these ideas are unformed, and will certainly benefit from deeper thought in relation to allied fields. In the spirit of this document they are offered for discussion, however, and for development or not as their merits may require.

Specific areas for research:

D1 – The role of excavation and analytical survey

As with all archaeological research, different sites and areas will demand different techniques dependent upon the nature of surviving remains and the questions being asked. But with most areas of archaeological research those decisions are taken within broad parameters determined by experience and know-how. For military archaeology of the recent past the parameters are not yet in place. Although a few excavations and detailed surveys of modern military sites have been completed (some purely to aid presentation) this lack of experience makes it more difficult to set well-thought out and meaningful archaeological conditions and goals for evaluation and fieldwork projects. Work is needed to determine what level of archaeological recording is sufficient or whether other evaluation techniques are appropriate for determining and interpreting what survives. A scoping study of work undertaken to date is needed to determine what was learnt from the various techniques employed on projects completed to date. Following that, and depending on the strength of its outcomes and conclusions, a wide ranging survey may be necessary, involving the excavation and survey of a few extant sites, involving structural recording, survey and excavation strategies, collection of oral accounts and archive accounts (where available) and a critique of the methods applied. This experience can then contribute to determining the parameters within which future research and evaluation and recording strategies are determined.

The activity of excavation or survey as an event may also be used to inspire media and personal interest and (re)create memories about a site's function and meaning; the intimate environment of an excavation could also provide the catalyst for dialogue amongst the groups once separated by the material culture they now join forces to understand. These thoughts are equally relevant for the many wartime and Cold War sites, including peace camps.

Following a scoping study, a survey is needed of the relevance and application of conventional archaeological methods and techniques to modern military sites and structures, perhaps involving well recorded and understood classes such as anti-aircraft sites

Figure 39   Former RAF Upper Heyford, Oxfordshire. This self-contained complex was constructed during the 1970s to accommodate armed F111 ground attack aircraft. Aircraft from the unit would have been amongst the first to respond to any Warsaw Pact attack (© Crown copyright NMR, 18537/18)

and/or coast batteries and a defended area in the first instance. The results, if worthwhile, might be published as required. Some experimentation with new techniques and ideas will be a useful next stage, especially for poorly understood sites and sites of discord. The value of cognitive mapping for example is worth exploring, to examine the influence of military sites on the local landscape and its inhabitants.

D2 – Social archaeology and interpretation of layout

In studying the archaeology of historic settlement sites and the remains of ships on the seabed, the spatial analysis of activities and areas is significant as is the meaning attached to these spaces by users. Logic suggests that this subject may not be worthwhile for modern sites for which detailed archives and oral testimonies exist. But there may be reasons to study spatial dimensions archaeologically, not

least to test that assumption. Artefact distributions within these sites may also provide a new dimension on the activities undertaken on site, and the discipline of soldiers, sailors and airmen. Where sites are available for excavation, do artefact distributions tell us anything useful about past activity beyond what we might glean from oral testimonials and archives? At a broader level, the spatial analysis of entire sites may be revealing, exploring the sight lines from key buildings and what these always embrace and what they always avoid; access lines around the site might also be significant. Are there comparisons between the principles of layout underpinning army camps, Royal Air Force stations, and naval bases for example?

Within buildings and ships (or submarines), the allocation of space between ranks and the provision of facilities may be worth studying archaeologically, as it could be between the sexes. The use of space in barracks for instance, and the trend from large open rooms to smaller more individual rooms; how

Figure 40   The married quarters at Biggin Hill, Kent. Typically for an inter-war station, these show the influence of earlier Garden City architecture (Photograph: Jeremy Lake)

housing estates reflect rank and grade creep, which for the Americans shows an increased tendency to reward technical personnel with rank, and with it an entitlement to a commensurate level of accommodation. Does grade creep in fact exist at all, or has this more to do with social changes within the services? The services have conducted their own research on private home ownership for example, and the effects on service personnel retention. Differences between the services may merit further research in this area: the Royal Naval personnel now generally occupy homes around their home port; army personnel move regularly between garrisons; while Royal Air Force staff often live and work in one area for their entire service career.

A research exercise exploring spatial analysis of modern military sites, and what additional information can be gleaned from the sites and their archaeology is a study worth undertaking, perhaps through postgraduate studies.

D3 – Oral history

Personal accounts are useful sources for documenting and interpreting social activities on military sites. But typically these are driven by explicitly social-historical agenda, and often omit asking the types of questions that might aid a fuller interpretation of a site's former use. Questions concerning, for example, use of space and discard

patterns. For the Cold War period oral-historical evidence is seen as one of the key methods for advancing understanding. In the interpretation of physical remains technical questions need to be asked relating to on-site maintenance and servicing procedures for various missile systems. Likewise little is known about safeguarding, handling and maintenance procedures surrounding nuclear weapons. Former personnel will be able to reflect on their attitudes to such weapons systems which would be unobtainable from other sources, though service personnel may now be reluctant to speak following recent prosecutions under the Official Secrets Act. The involvement and cooperation of MoD and other government departments may prove helpful in advising on the status of information. Beyond these technical questions, other issues such as daily routines, travel, messing and leisure could be explored. Testimonies may also be used to challenge official histories, which often put a positive gloss on things and impose a logical development process on, for example, weapons procurement programmes, where the reality is often more confused. Detailed information does exist on weapons procurement programmes, but most remains classified.

Where sites exist and former servicemen or occupants can be traced, an archaeologically driven research programme investigating use of space could be considered. One example might be those few PoW camps where former prisoners

Figure 41   A 6 inch gun emplacement at Beacon Hill Battery, Harwich, Essex (Photograph: Roger J C Thomas)

stayed on and continue to live locally. Again this could form the subject of a research degree, embracing social historical and archaeological objectives. For the Cold War a specific project, with technological and social historical objectives is needed. Many recently redundant USAF airfields have dedicated websites, where veterans post reminiscences about former postings; these are a valuable resource as are warship reunion groups and Regimental historians.

D4 – 'Personality' of military areas

It is recognised that areas occupied by military units will often develop a character or personality that reflects that presence, and the longer-lasting the military presence, or the more intense it is, the stronger that personality can become. This is one of the main ways in which military heritage can contribute to the new and emerging characterisation agenda (Fairclough *et al* 2002).

A social historical survey which examines these issues is planned for East Anglia, exploring the influence of the World War II and Cold War presence of the USAF on local communities and the landscape. A further such study could usefully take in Greenham Common and the Newbury area. Much useful work has been done on the military 'personality' of Orkney and specifically the area around Scapa Flow by RCAHMS. A characterisation project to explore the English landscape, 1946–2000 is currently in preparation.

E – Management principles and frameworks

Objective: to ensure appropriate and effective measures and procedures are in place to preserve and manage twentieth-century military resources alongside those of the more distant past.

As a comparatively new addition to the heritage further work is needed to establish modern military remains within the context of conservation practice and philosophy. The main areas where progress is needed are record keeping, adopting a common typology and terminology, and ensuring these sites sit alongside others in management frameworks.

Figure 42   A World War II re-enactment group in action at Tilbury Fort, Essex (Photograph: John Schofield)

Specific areas for research/progress:

E1 – Terminology

There is a need for a thesaurus (to include seafaring and maritime terms), based on contemporary and agreed terminology. A published (and illustrated) thesaurus and handbook, made widely available – including via the Internet – to local authorities and other interest groups, is a fundamental first step.

One product arising from the Defence of Britain and other related projects (eg MPPs' work) should be an online (and hard copy) thesaurus of twentieth-century military remains in Britain. It is hoped the national agencies, alongside Association of Local Government Archaeological Officers (ALGAO) and the Fortress Studies Group, for example, could oversee and sponsor this, but within the context of related projects. Specialists should be consulted for topics such as the Cold War, for which primary and secondary source material may not always be available.

E2 – Preservation of archives

Archives relating to modern warfare, and which are significant for archaeological, historical and social historical research, survive in numerous locations, and under various conditions. Some archives held at the National Archives, and those most useful for archaeological research (eg site plans), may be considered by the National Archives to be too specialised, though the resource may be too large for other depositories (eg National Monument Record). Military museums and libraries often have significant holdings. MoD also holds valuable archives (such as the vital Air Ministry Registered Drawings and terrier books), often uncatalogued and held locally, while other sources include the Imperial War Museum, the Royal Air Force Museum at Hendon and local record offices such as that in Winchester which holds an important collection of Royal Navy records relating to shore facilities. Contemporary photographs, many taken unofficially, can provide valuable information on the life of redundant defence facilities.

Meetings should be arranged involving appropriate staff from all organisations which hold historic military archives relating to the function and form of specific military sites, with a view to determining priorities for retention, and agreeing suitable locations. This could be set up initially by the National Monument Record, with the National Archives, Imperial War Museum and MoD as key partners. One priority is to catalogue Air Ministry Drawings, vital for archaeological and architectural research on airfields.

Figure 43   Surviving military sites are often unspectacular, and sometimes baffling to the untrained eye, but to local residents they can represent important reminders and contribute much to the sense of place. Here a spigot mortar emplacement survives at Goodwick, Pembrokeshire (Photograph: Roger J C Thomas)

E3 – Record keeping

Ensure, through the appropriate channels, that Defence of Britain Project data, alongside those arising from MPP, the National Mapping Programme (NMP) and sites separately entered onto SMRs and the National Monument Record can all be accessed in the same places (and notably SMRs and the NMR), whether by developers or researchers.

The mechanisms for achieving this objective are already in place, though further discussions with ALGAO, NMR staff and NMP will ensure this continues to be the case, and that new initiatives are engaged at an early stage.

E4 – Protection of sites

To ensure the completion of reviews and work programmes that result in appropriate sites having statutory protection, especially in the cases of the more vulnerable and rare monument classes (eg bombing decoys and D-Day sites). There is also a need to ensure that sites continue to be fully recorded on both the NMR and locally held SMRs, meaning that they can be treated as appropriate through the development control process. Records should include both known surviving sites, and those documented in the MPP reports, where some significant buried remains may exist. Finally, it is imperative that measures are put in place to fully record documented aircraft crash sites on SMRs and the NMR, in addition to MoD maintaining their own records. For Anti-invasion defences a catalogue or list of defence landscapes may be appropriate, following the English Heritage defence areas project.

There needs to be incorporation of military sites in Designation Team work programmes, as well as comparable programmes elsewhere. Liaison with NMR staff and with ALGAO should ensure that the need to maintain adequate records is realised. The crash site work also requires continued liaison with MoD, and the idea of producing a list of defence landscapes will require further internal negotiation and discussion including with the DCMS. The Defended Areas project is already underway and the results of this will eventually feed into local planning agenda, and designation programmes.

E5 – Management, presentation and interpretation

Research leading to the publication of best practice

guidelines for the management, presentation and interpretation of military heritage sites, on land and underwater, would be useful for curators, conservators and practitioners. With so many of these sites now held and managed by national heritage agencies, local groups and trusts, and – increasingly – individual owners, such guidance is timely, and would have international appeal. This should be both practical (for example, on earthwork conservation, and determining, for example, that only the guns intended for a site should be presented there) and philosophical (enabling managers to decide whether to present the past as it was or in some diluted form). The Vimy Declaration (currently in draft) on conserving battlefield terrain is a model of what can be achieved in this field (see http://www2.cr.nps.gov/abpp/terraincharter.htm). These guidelines could also usefully address the role contemporary art plays in interpreting military sites. With the experience of managing sites like Dover Castle and Fort George, this best practice guidance is something the national heritage agencies should consider commissioning. Further 'best practice' guidelines in managing specific classes of monument (marine, wall art, airfields, for example) should be produced as needs arise.

Initially discussions should be held to address the market for such guidelines, their content and scope, potential authors, and funding streams.

F – Articulation, co-ordination and publication

Objective: to co-ordinate the objectives and aspirations of the many groups, individuals and specialised archaeological organisations and agencies who seek to develop an understanding of the scale and logic of militarisation in Britain throughout the twentieth century, for the purposes of better working practices, and improved understanding and awareness. This should be achieved through cooperation and networking, the university sector and voluntary and local organisations being key players.

Specific areas for research:

F1 – Local level

Opportunities should be sought for local studies within the terms of this discussion document and the national programmes of research undertaken to date. Defence heritage, perhaps more than any other subject, lends itself to this approach given the need to tap oral-historical evidence, local archives and records, past news coverage, and field remains. Here is the opportunity for community archaeology, for engaging parish councils, local history groups and schools in a wide field of study that has national significance and relevance. As an example, anti-invasion defences in particular provide an opportunity to examine the close relationships that existed in the minds of military planners between defence and militarisation and the

natural and built environment. Defended areas, where these survive now much as they were in 1940, provide opportunities to study this relationship in terms of military tactics and strategies of defence and counter-attack; also in terms of the impact the military presence had on the local community. This consideration of defended areas or military landscapes has potential for public education and enjoyment, as well as having a role within the national curriculum.

By promoting this subject through publications, talks etc, and through teacher training days – of the type organised by English Heritage SE Region – local studies will emerge. Professional archaeologists and curators should encourage such initiatives and guide them in terms of advice, provision of records and quality control input.

F2 – Regional level

Local studies feed inevitably into the regional picture, while equally regional research agenda can determine priorities and programmes for local groups and communities to pursue. In terms of public understanding at this wider landscape scale, published 'trails' can reveal the extent to which parts of Britain have become militarised areas. The naval presence can be strongly felt around Portsmouth and Gosport, for example; the army in areas such as Aldershot and East Hampshire, and the RAF (and USAF) in East Anglia. An historic airfields trail has been published for Lincolnshire to address this issue and meet the growing demand for information. A 'bomber landscape' project has been suggested for Lincolnshire, to assess the impact of the military presence and personnel on the landscape and its inhabitants.

As at the local scale (above), access to these militarised areas, and to information about them, is in growing demand and published leaflets and books are worthwhile in promoting access and understanding. Local authority museum services are already playing a major role in this: two exhibitions to increase awareness of the resource in Dumfries and Galloway took place in 2002. Curators and national heritage agency staff should ensure that modern military heritage is not ignored in emerging regional research agenda (Glazebrook 1997; Brown and Glazebrook 2000 for examples of a regional research framework accommodating recent military heritage).

F3 – National level

There are significant issues in the developing field of defence heritage that require wide appreciation within and beyond the profession, and crucially at a national level. These include a need to appreciate the range of resources available, for example at the National Archives and SMRs, and the requirement that research be fed back to SMRs, in order that results can be available to others, and accommodated within the planning system. There is also the related need for a coherent national (perhaps UK-wide) conservation strategy for these sites, preferably in the form of a published statement to help the national agencies as well as local authorities and government to achieve a degree of consistency. Coordination of university projects and research interests, with those undertaken by amateur archaeologists and historians, the national heritage agencies and others is also a necessity and, for this, an Internet discussion group may suffice in the short term at least. The example of English Heritage's Military and Naval Strategy Group could usefully be followed elsewhere, though linkages to other related strategy groups (eg urban, industrial, rural and maritime) need to be developed for this system to be truly effective. In fact the whole subject of recent military remains should in time become integrated within the wider field of twentieth-century heritage. Finally, the national heritage agencies and specialist groups have a significant role in providing appropriate training for staff and volunteers at a local level, bringing the subject matter to a wider audience, promoting best practice, and ensuring new research is widely disseminated for the benefit of all.

Discussions are needed at an early stage on each of these objectives. The 'National Conservation Strategy' can be drafted as a subsequent document to this, while the others – and any related objectives – can be achieved by establishing an Internet discussion group, perhaps building on groups established by the CBA, and through the Defence of Britain Project. With the subject still in its infancy, training must remain a high priority. A short course on Oral History in Archaeological Practice, to provide training in gathering, recording and interpreting oral-historical evidence, is being offered by Bournemouth University.

F4 – International level

Defence heritage studies have an international context which provides the potential for exchange of information and expertise, and developing understanding of the subject at a wider geographical level. For the Cold War, for example, difference in the plan form of comparable sites from East and West may aid interpretation, as might the study of facilities relating to the production, testing, storage and use of nuclear weapons. The different rituals connected with these weapons systems, and their archaeological manifestation would be of interest. There is much potential here, for examining many topics of both world wars (eg comparing anti-invasion defences and defence strategies in Germany and Britain) and the Cold War. The impact of frontiers and boundaries/barriers is another subject that has wide geographical relevance.

Staff engaged with this subject at a strategic level should keep abreast of opportunities, funding streams and priorities promoted, for example, by the European Union. A Cold War European legacies project is already under consideration.

Part 3: The way ahead

What follows is a summary of the agenda described above, and a prioritisation for the topics listed. However, this document has a limited shelf-life. In fact it is current only at the time of writing, the subject developing rapidly as thoughts continue to evolve and projects are completed. To be realistic, a 3–4 year shelf life is felt appropriate for this initial document, with a redraft therefore due in 2007 at the latest. It is also anticipated that following the publication of this discussion document, a series of Implementation Plans will be produced, perhaps one for each of the home countries (though dependant on staff time and resources). The English plan will be prepared shortly.

These Implementation Plans will confirm the priorities for research, as well as suggesting outline costs and timetables for completing the work. They will also set out procedures for monitoring progress. These Implementation Plans will have a shorter shelf-life, with a 2–3 year cycle envisaged. For the initial Implementation Plan for England, projects currently underway or earmarked will be identified, as will projects likely to fall within the next phase of work. It is likely that over time basic research into site distributions and survival will give way to projects that are more concerned with conservation management and dissemination.

Table 1   Summary of frameworks and objectives, with indicative timescale

Framework	Objectives [with themes]	Immediate	Short term	Medium term
A Improve understanding of the built resource	A1 D-Day preparations and support [3,4]		+	
	A2 Camps (general) [3]		+	
	A3 Cold War [1–3]			+
	A4 Civil infrastructure [3]		+	
	A5 Searchlight emplacements and barrage balloon sites [4]			+
	A6 Internal security and aid to the Civil Power [4]		+	
	A7 Intelligence infrastructure [3,4]		+	
	A8 Air navigation aids, 1939–45 [3,4]		+	
	A9 BBC wartime and Cold War broadcasting [3]		+	
	A10 Royal dockyards [3]			+
	A11 Warships and submarines [3,4]		+	
	A12 Z batteries [4]			+
	A13 Overview [1–4]			+
B Improve understanding of surviving resources	B1 Phase 2 surveys for site types investigated under A, above [3,4]		+	
	B2 World War 1 [1–4]		+	
	B3 Submerged archaeology [1,4]			+
	B4 Anti-invasion defences [4]		+	+
	B5 Anti-aircraft, decoys, radar and coast artillery [4]		+	
	B6 Utilities and communications [4]		+	
	B7 Training areas [1]		+	
	B8 PoW camps [3]	+		
	B9 Territorial Army [3, 4]	+		
	B10 Civil defence [4]		+	+
	B11 Hospitals [3]		+	
	B12 Cold War [1–4]		+	
	B13 Wall art [2–4]	+		
	B14 Countryside at War [1–5]			+
C Pressures and perceptions	C1 Crash site excavation and loss of records [4]	+		
	C2 Stability and conservation [1–4]	+	+	
	C3 Changing attitudes [1–5]		+	+
	C4 Social commemoration of warfare [1–5]		+	

Table 1 **Summary of frameworks and objectives, with indicative timescale**

Framework	Objectives [with themes]	Immediate	Short term	Medium term
D Methodologies	D1 The role of excavation and analytical survey [1–4]	+		
	D2 Social archaeology and interpretation of layout [1–4]		+	
	D3 Oral history [1–5]	+	+	
	D4 Personality of military areas [1,4]		+	+
E Management principles and frameworks	E1 Terminology [1–4]	+		
	E2 Preservation of archives [1–4]	+		
	E3 Record keeping [1–5]	+		
	E4 Protection and management of sites [1–5]	+		
	E5 Presentation/interpretation [1–5]		+	
F Articulation, coordination and publication	F1 Local [1–5]	+		
	F2 Regional [1–5]	+		
	F3 National [1–5]	+		
	F4 International [1–5]	+		

Immediate refers to the next 1–2 years, short term, the next 3–4 years, and medium term is beyond that.

Conclusion

Over the past 10–15 years, recent military heritage has attained a credibility and support base, for example amongst curators, contractors and staff of the academy, confirming its place as a serious and worthwhile pursuit. Specifically during this period there have been numerous research programmes undertaken in the name of informed conservation and to promote public enjoyment and awareness of this previously little-understood aspect of cultural heritage. Serious debate has also concerned the validity of this heritage: is it something to be preserved and retained, or is it best ignored and hidden from view? Or does it have even greater relevance at times when war once again seems likely? Will monuments of the 'Troubles' in time reinforce the peace process in Northern Ireland, for example, or could their presence compromise moves towards peace? And what of motivations: are these sites retained as memorials to the fallen, or do they simply represent a part of our cultural heritage; part of the story? So, alongside the recording and research into the archaeology of this period is now occurring serious debate about motives and meaning (eg Schofield *et al*

2002; Virilio and Lotringer 1997), indicative of a healthy discipline and suggestive of a productive and healthy future. Finally, underpinning all of these recent developments has been the valuable work undertaken by amateur researchers and enthusiasts over the past two or three decades culminating most recently in the completion of the Defence of Britain Project and ultimately virtually all of the research outlined in Part 1 of this document.

A significant threshold has now therefore been reached. A large enthusiastic and committed constituency has emerged embracing archaeologists, historians, archivists, sociologists, anthropologists and – significantly – service personnel (whose advice is invaluable if we are to get it right, especially for the recent periods where documents may remain closed); community support exists as evidenced by book sales, viewing figures for television programmes, and the clear messages of support for characterisation, recording and designation programmes much valuable research has been undertaken providing for many of the major categories and classes of site,

Figure 44   The fence enclosing the Ground Launched Cruise Missiles Alert and Maintenance Area (GAMA) at Greenham Common, and separating the military estate from one of the peace camps that surrounded the former airbase. Cuts to the fence, as here, can still be seen (Photograph: John Schofield)

Figure 45   Anti-motor torpedo boats emplacements added to a nineteenth-century fortification on the Isle of Grain, Kent (Photograph: John Schofield)

Figure 46   The Maunsell Sea Forts, anti-aircraft emplacements built in the Thames Estuary, and reused by pirate radio stations in the 1970s (Photograph: John Schofield)

information on site location, modern survival, typology and operational considerations; and now a research framework has been produced drawing this work together and identifying where further work is needed and what priorities exist given the current position on threat, including disposals, re-use and regeneration. It would be easy following these considerable efforts of the past decade or so to now consider this work done; to recognise this as the beginning of the end. But research frameworks should never be that, and certainly not for a subject still so new (including of course by definition). Rather, this stage represents the end of a very productive and encouraging period in which understanding has advanced significantly. The foundations are in place from which to build a mature and integrated sub-discipline which has direct and clear relevance in the modern world. But to avoid complacency, we should ensure those of us engaged in the subject remain abreast of developments, and be prepared to update or reassess our understanding of what may appear fixed and complete, as change occurs, or as new information (notably from classified sources) becomes available.

Finally, 'Total War' engaging everywhere and everybody is a characterising feature of the period covered by this document, and as such the separation of war from other aspects of twentieth-century culture and heritage is a boundary we should eventually seek to remove. As understanding improves, and the questions in subsequent military and other twentieth century heritage frameworks documents become more focussed, our approach to all aspects of twentieth-century culture should be drawn together into a more integrated survey of the period. But these are disciplinary boundaries that should only be dismantled when the time is right.

The twentieth century was a period of immense cultural, social and scientific change, and arguably the worst and most horrific in human history. Eric Hobsbawm famously described it as an 'age of extremes'. But it is a past not to be hidden away, however difficult it might be, though the challenges we face in recording and interpreting it will sometimes create tensions. This military framework is the first attempt to determine how research is taken forward in a constructive, effective and – where appropriate – careful and sensitive way. If it creates the opportunity for exciting and relevant inter-disciplinary research programmes of the type described in this report, it will have been a success.

Useful addresses

English Heritage
23 Savile Row
London W1S 2ET
020 7973 3000
www.english-heritage.org.uk (Includes pages on
recent military heritage, and links to relevant free
publications)

National Monuments Record Centre
Kemble Drive
Swindon SN2 2GZ
01793 414600

Historic Scotland
Longmore House
Salisbury Place
Edinburgh EH9 1SH
0131 668 8638
www.historic-scotland.gov.uk

Cadw: Welsh Historic Monuments
Crown Building
Cathays Park
Cardiff CF2 1UY
029 2082 5449
www.cadw.wales.gov.uk

Environment and Heritage Service, Northern Ireland
5/33 Hill Street
Belfast BT1 2LA
01232 235000
www.ehsni.gov.uk

Manx National Heritage
Douglas
Isle of Man IM1 3LY
01624 648000
www.gov.im/mnh/

Council for British Archaeology
Bowes Morrell House
111 Walmgate
York YO1 9WA
01904 671417
www.britarch.ac.uk/projects/dob (The CBA's Defence
of Britain Project website, with news, information and
the project review)
http://ads/ahds.ac.uk/catalogue/resources.html?dob
(The Defence of Britain Project database, that can be
searched and interrogated online)

Association of Local Government Archaeological
Officers (ALGAO)
C/o Heritage Consultation Group
Planning Division
Essex County Council
County Hall
Chelmsford CM1 1LF
01245 437676
www.algao.org.uk

Imperial War Museum
Lambeth Road
London SE1 6HZ
020 7416 5000
www.iwm.org.uk

National Archives
Ruskin Avenue
Kew
Richmond
Surrey TW9 4DU
020 8876 3444
www.nationalarchives.gov.uk

Fortress Study Group
W H Clements
6 Lanark Place
London W9 1BS

Friends of War Memorials
Lower Belgrave Street
London SW1W OLA
www.war-memorials.com

British Aviation Archaeological Council
BAAC Honorary Secretary
Spring View
Kenilworth
Warwickshire CV8 2JS

Subterranea Britannica
www.subbrit.org.uk (Subterranea Britannica's website, containing much information on predominantly
underground and Cold War military sites). See
www.subbrit.org.uk/rsg/index.shtml for their Cold
War Research Study Group.

Pillbox Study Group
www.pillbox-study-group.org.uk. The Pillbox Study
Group is a forum for those interested in World War II
anti-invasion defences, with a website and quarterly
journal.

References

Anderton, M, 2000 Assessment of surviving Cold War Rotor Stations. MPP Report, Internally circ.

Anderton, M, and Schofield, J, 1999 Anti-aircraft gunsites – then and now. *Conservation Bulletin* **36**, 11–13.

Barnes, J, 2002 *Second World War defences in Kent: information for teachers*. English Heritage.

Betts, R A, 1996 *The Royal Air Force Construction Service, 1939–1945*. Ware: Airfield Research Publishing.

Blake, R, Hodgson, M, and Taylor, B, 1984 *The Airfields of Lincolnshire since 1912*. Leicester.

Bragard, P, Termote, J, and Williams, J, 1999 *Walking the Walls: Historic town defences in Kent, Cote d'Opale and West Flanders*. Kent County Council, le Syndicate Mixte de la Cote d'Opale and the Province of West Flanders.

Brown, N, and Glazebrook, J, (ed) 2000 Research and Archaeology: a Framework for the Eastern Counties, 2. Research agenda and strategy. *East Anglian Archaeology*, Occasional Paper 8. Scole Archaeological Committee.

Cocroft, W D, 2000 *Dangerous Energy: The archaeology of gunpowder and military explosives manufacture*. English Heritage.

Cocroft, W D, 2001 *Cold War Monuments: an assessment by the Monuments Protection Programme*. English Heritage. [Available in CD format.]

Cocroft, W D, 2003 The Cold War: What to preserve and why. *Conservation Bulletin* **44**, 40–42.

Cocroft, W D, and Schofield, J, 2003 Images of the Cold War: Combat art. *Conservation Bulletin* **44**, 43–4.

Cocroft W D, and Thomas, R J C, 2003 *Cold War: building for nuclear confrontation, 1946–89*. English Heritage.

DCMS, 2002 *A Force for our Future*. London: The Stationery Office.

Dobinson, C S, 1995 Twentieth Century Fortifications in England, Vol III: Bombing decoys of World War II. CBA/English Heritage. [Available at NMR and SMRs.]

Dobinson, C S, 1996a Twentieth Century Fortifications in England, Vol IV: Operation Diver. CBA/English Heritage. [Available at NMR and SMRs.]

Dobinson, C S, 1996b Twentieth Century Fortifications in England, Vol I: Anti-aircraft artillery. 1914–46. CBA/English Heritage. [Available at NMR and SMRs.]

Dobinson, C S, 1996c Twentieth Century Fortifications in England, Vol II: Anti-invasion defences of World War II. CBA/English Heritage. [Available at NMR and SMRs.]

Dobinson, C S, 1996d Twentieth Century Fortifications in England, Vol V: Operation Overlord Preparatory Sites. CBA/English Heritage. [Available at NMR and SMRs.]

Dobinson, C S, 1997 Twentieth Century Fortifications in England, Vol IX: Airfield themes. CBA/English Heritage. [Available at NMR and SMRs.]

Dobinson, C S, 1998a Twentieth Century Fortifications in England, Vol X: Airfield defences in World War II. CBA/English Heritage. [Available at NMR and SMRs.]

Dobinson, C S, 1998b Twentieth Century Fortifications in England, Vol XI: The Cold War. CBA/English Heritage. [Available at NMR and SMRs.]

Dobinson, C S, 1999a Twentieth Century Fortifications in England, Vol VII: Acoustics and Radar. CBA/English Heritage. [Available at NMR and SMRs.]

Dobinson, C S, 1999b Twentieth Century Fortifications in England, Vol VI: Coast artillery, 1900–56. CBA/English Heritage. [Available at NMR and SMRs.]

Dobinson, C S, 1999c Twentieth Century Fortifications in England. Supporting Paper AA/1: Searchlight sites of World War II: a sample list. CBA/English Heritage.

Dobinson, C S, 1999d Twentieth Century Fortifications in England, Vol VIII: Civil Defence in World War II. CBA/English Heritage. [Available at NMR and SMRs.]

Dobinson, C S, 2000a Twentieth Century Fortifications in England. Supplementary Study. Experimental and Training Sites: an annotated handlist. CBA/English Heritage.

Dobinson, C S, 2000b *Fields of Deception: Britain's bombing decoys of the Second World War*. London: Methuen.

Dobinson, C S, 2001 *AA Command: Britain's anti-aircraft defences of the Second World War*. London: Methuen.

Dobinson, C S, Lake J and Schofield, J, 1997 Monuments of War: defining England's twentieth century defence heritage. *Antiquity* **71**, 288–99.

Dolff-Bonekaemper, G, nd Sites of historical significance and sites of discord: historic monuments as a tool for discussing conflict in Europe in *Forward Planning: the function of cultural heritage in a changing Europe*. Council of Europe.

Douet, J, 1998 *British Barracks 1600–1914: their architecture and role in society*. English Heritage.

Doughty, M (ed), 1994 *Hampshire and D-Day*. Hampshire Books.

Edgerton, D, 1991 *England and the Aeroplane. An Essay on a Militant and Technological Nation*. London: Macmillan.

English Heritage, 1998 *Monuments of war: the evaluation, recording and management of twentieth-century military sites*. English Heritage.

English Heritage, 2000a *Twentieth-century military sites*. English Heritage.

English Heritage, 2000b *Power of Place*. English Heritage.

English Heritage, 2001 Buildings of the Radio Electronics Industry in Essex. Typescript report.

English Heritage, 2002 *Military Aircraft Crash Sites: archaeological guidance on their significance and future management*. English Heritage.

English Heritage, 2003 *Management guidelines for military aviation sites*. English Heritage.

Foot, W, 2003 Public archaeology: defended areas of World War II. *Conservation Bulletin* **44**, 8–11.

Fairclough, G, and Rippon, S (eds) 2002 Europe's Cultural Landscape: archaeologists and the management of change, *EAC Occasional Paper* **2**.

Francis, P, 1996 *British Military Airfield Architecture*. Sparkford: Patrick Stephens.

Francis, P, nd A survey of WW2 airfield defences in England. An unpublished report for English Heritage.

Gander, T, 1979 *Military archaeology: a collector's guide to 20th century war relics*. Cambridge: Patrick Stevens.

Gillett, S, 1999 The Aircraft Industry in Avon and Gloucestershire. MSc thesis, Ironbridge Institute.

Glazebrook, J (ed), 1997 Research and Archaeology: a Framework for the Eastern Counties, 1: Resource assessment, *East Anglian Archaeology*, Occasional Paper **3**. Scole Archaeological Committee.

Going, C, 2002 A neglected asset. German aerial photography of the Second World War period, in R Bewley and W Raczkowski, (eds), *Aerial Photography: developing future practice*, pp. 23–30. IOS Press: NATO Science Series 1: Life and Behavioural Sciences Vol. **337**.

Hellen, J A, 1999 Temporary settlements and transient populations. The legacy of Britain's prisoner of war camps. *Erdkunde* (Archive for Scientific Geography) **53(3)**, 191–211.

Hoare, P, 2001 *Spike Island: the memory of a military hospital*. London: Harper Collins.

Holyoak, V, 2001 Airfields as battlefields, aircraft as an archaeological resource: British military aviation in the first half of the twentieth century, in P Freeman and A Pollard (eds), *Fields of Conflict: Progress and Prospect in Battlefield Archaeology*, 253–64. BAR International Series **958**.

Holyoak, V, 2002 Out of the blue: assessing military aircraft crash sites in England, 1912–45, *Antiquity* **76**, 657–63.

Hughes, M, 1992 The archaeology of D-Day: the remains at Shore Point, Lepe. *Fortress* **15**, 57–62.

James, N, 2002 The Cold War, *Antiquity* **76**, 664–6.

Jarman, N, 2002 Troubling remnants: dealing with the remains of conflict in Northern Ireland in J Schofield, W G Johnson, and C M Beck (eds), *Matériel Culture: the archaeology of twentieth-century conflict*, 281–95. Routledge. One World Archaeology **44**.

Lacey, C, 2002 Southampton at War – protecting the public: the survival and management of Second World War and Cold War civil defence structures in a local urban context. Unpublished undergraduate dissertation: University of Southampton.

Lake, J, 2000 Thematic Survey of Military Aviation Sites and Structures. Unpublished report, English Heritage Thematic Listing Programme.

Lake, J, 2002 Historic airfields: evaluation and conservation in J Schofield *et al* (eds), *Matériel Culture: the archaeology of twentieth-century conflict*, 172–88. Routledge. One World Archaeology **44**.

Lake, J, and Schofield, J, 2000 Conservation and the Battle of Britain, in P Addison, and J Crang (eds), *The Burning Blue: a new history of the Battle of Britain*, 229–42. London: Pimlico.

Lloyd, V, 2001 *Richmond Castle – conscientious objection and the Richmond Sixteen: information for teachers*. English Heritage.

Lowry, B, (ed), 1995 *20th Century Defences in Britain: an introductory guide*. Handbook of the Defence of Britain Project. CBA.

Martin, R, 2002 King's Standing Transmitter Station, Crowborough, *Industrial Archaeology Review XXIV*, 91–102.

McOmish, D, Field, D, and Brown, G, 2001 *The Field Archaeology of Salisbury Plain Training Area*. English Heritage.

Nash, F, 2002 World War Two defences in Essex, *After the Battle* **116**, 30–35.

Oxley, I, 2002 Scapa Flow and the protection and management of Scotland's historic military shipwrecks, *Antiquity* **76** (293), 862–8.

RCAHMS, 1999a Catalogue of the Luftwaffe Photographs in the National Monuments Record of Scotland; Scotland from the Air, 1939–49, **Vol 1**. RCAHMS.

RCAHMS 1999b Catalogue of the RAF World War II Photographs in the National Monuments Record of Scotland; Scotland from the Air, 1939–49, **Vol 2**. RCAHMS.

RCHME, 1994a Bowaters Farm survey.

RCHME, 1994b The Royal Gunpowder Factory, Waltham Abbey, Essex: an RCHME Survey. Typescript report.

Read, P, 1996 *Returning to nothing: the meaning of lost places*. Cambridge: Cambridge University Press.

Redfern, N, 1998a Twentieth Century Fortifications in the United Kingdom: Vol. 1: Introduction and Sources. York: CBA.

Redfern, N, 1998b Twentieth Century Fortifications in the United Kingdom: Vol. 2: Site gazetteers: Wales. York: CBA.

Redfern, N, 1998c Twentieth Century Fortifications in the United Kingdom: Vol. 3: Site gazetteers: Scotland. York: CBA.

Redfern, N, 1998d Twentieth Century Fortifications in the United Kingdom: Vol. 4: Site gazetteers: Northern Ireland. York: CBA.

Richardson, H, (ed), 1998 *English Hospitals 1660–1948: a survey of their architecture and design*. London: RCHME.

Saunders, A, 1989 *Fortress Britain: artillery fortification in the British Isles and Ireland*. Beaufort.

Schofield, J, 2001 D-Day sites in England: an assessment, *Antiquity* **75** (287): 77–83.

Schofield, J, 2002a The role of aerial photographs in national strategic programmes: assessing recent military sites in England, in R Bewley and W Raczkowski (eds), *Aerial Photography: developing future practice*, pp. 269–82. IOS Press: NATO Science Series 1: Life and Behavioural Sciences Vol **337**.

Schofield, J, 2002b Monuments and the memories of war: motivations for preserving military sites in England, in J Schofield *et al* (eds), *Matériel Culture: the archaeology of twentieth century conflict*, 143–58. Routledge. One World Archaeology **44**.

Schofield, J, In press Aftermath: materiality on the home front, 1914–2001 in N Saunders (ed), *Materialities of Conflict: Anthropology and the Great War, 1914–2001*. London: Routledge.

Schofield, J, and Anderton, M, 2000. The queer archaeology of Green Gate: interpreting contested space at Greenham Common Airbase. *World Archaeology* 32, 236–51.

Schofield, J, Johnson, W G, and Beck, C M (eds), 2002. *Matériel Culture: the archaeology of twentieth century conflict*. Routledge. One World Archaeology **44**.

Schofield, J, Webster, C J, and Anderton, M J, 2001 *Second World War Remains on Black Down: a reinterpretation*. Somerset Archaeology and Natural History, 1998. 271–86.

Searle, A, 1995 PLUTO: *Pipeline under the ocean*. Shanklin: Shanklin Chine.

Smith, V, 2001 *Front-line Kent. Defence against invasion from 1400 to the Cold War*. Maidstone: Buckland Press.

Thomas, R J C, 1994 Survey of 19th and 20th century military buildings of Pembrokeshire. Unpublished report.

Thomas, R J C, 2003 PoW camps: What survives and where. *Conservation Bulletin* **44**, 18–21.

Tivers, J, 1999 'The Home of the British Army': the iconic construction of military defence landscapes, *Landscape Research* **24:3**, 303–19.

Virilio, P, and Lotringer, S, 1997 *Pure War* (Revised edition). New York: Semiotext(e).

Wainwright, A, 1996 Orford Ness, in D Morgan Evans, P Salway and D Thackray (eds), *The Remains of Distant Times*, 198–210. Boydell.

Wessex Archaeology, 2000a *Assessment of the Royal Dockyards at Portsmouth*. Wessex Archaeology.

Wessex Archaeology, 2000b *Assessment of the Royal Dockyards at Devonport*. Wessex Archaeology.

Wills, H, 1985 *Pillboxes: A Study of UK Defences 1940*. Leo Cooper.